Molecular Design and Synthesis of Novel Energetic Compounds

Guest Editor

Jianguo Zhang

MDPI

Basel • Beijing • Wuhan • Barcelona • Belgrade • Novi Sad • Cluj • Manchester

Guest Editor
Jianguo Zhang
State Key Laboratory of
Explosion Science
and Technology
Beijing Institute of Technology
Beijing
China

Editorial Office
MDPI AG
Grosspeteranlage 5
4052 Basel, Switzerland

This is a reprint of the Special Issue, published open access by the journal *Molecules* (ISSN 1420-3049), freely accessible at: https://www.mdpi.com/journal/molecules/special_issues/8V6SF77453.

For citation purposes, cite each article independently as indicated on the article page online and as indicated below:

Lastname, A.A.; Lastname, B.B. Article Title. *Journal Name* **Year**, *Volume Number*, Page Range.

ISBN 978-3-7258-5657-2 (Hbk)
ISBN 978-3-7258-5658-9 (PDF)
https://doi.org/10.3390/books978-3-7258-5658-9

Contents

About the Editor

Jianguo Zhang

Jianguo Zhang is a Professor in the State Key Laboratory of Explosion Science and Technology at the Beijing Institute of Technology (BIT). He obtained his Ph.D. degree in applied chemistry from BIT and has been a faculty number at BIT since 2000. His research interests include energetic materials and energy storage devices. He has co-authored over 300 research papers and filed more than 60 patents.

molecules

MDPI

Article

Development of Melt-Castable Explosive: Targeted Synthesis of 3,5-Dinitro-4-Methylnitramino-1-Methylpyrazole and Functional Derivatization of Key Intermediates

Elena Reinhardt [1], Lukas Bauer [1], Antonia H. Stadler [1], Henrik R. Wilke [1], Arthur Delage [2], Jörg Stierstorfer [1,*] and Thomas M. Klapötke [1,*]

1 Department of Chemistry, Ludwig-Maximilians-University of Munich, Butenandtstr. 5–13, 81377 Munich, Germany; elrech@cup.uni-muenchen.de (E.R.); luauch@cup.uni-muenchen.de (L.B.)

2 Eurenco, 84700 Sorgues, France; a.delage@eurenco.com

* Correspondence: jstch@cup.uni-muenchen.de (J.S.); tmk@cup.uni-muenchen.de (T.M.K.)

Abstract: The problems associated with TNT necessitate the development of novel melt-castable compounds with melting points between 70 and 120 °C, a crucial endeavor in the field of energetic materials. This study introduces a promising melt-castable explosive based on nitropyrazole, whose melt-castable properties were achieved by the introduction of methyl groups. The synthesis of 3,5-dinitro-4-methylnitramino-1-methylpyrazole involves a three-step process starting from 3,5-dinitro-4-chloropyrazole, including substitution, nitration, and methylation reactions. Additionally, two alternative synthesis routes and six energetic salts were examined. Structural elucidation was conducted using conventional methods such as NMR, IR, and X-ray, while the energetic properties of the compound, including thermal behavior, sensitivities, and theoretical performance, were investigated. Also, compatibility with common explosives was investigated, the experimental enthalpy of formation by bomb calorimetry was determined, and an SSRT test was performed. Furthermore, the melt-cast explosive underwent an Ames test in order to assess its toxicity.

Keywords: melt-castable explosives; pyrazole; compatibility; bomb calorimetry; SSRT; toxicity

1. Introduction

The class of industrially used melt-castable energetic materials consists of only a few compounds, most prominently trinitrotoluene (TNT), yet they are of fundamental importance in energetic material military and industrial applications including warheads, artillery shells, and civilian demolition charges [1,2]. TNT remains the industry standard, forming the basis of key formulations such as Composition B (RDX, TNT), Octol (HMX, TNT), and Pentolite (PETN, TNT) (Figure 1) [1,3]. However, its synthesis generates undesired asymmetrical TNT derivatives that must be removed [4]. This is typically achieved through washing with an aqueous sodium sulfite solution, known as the Sellite process, which produces large volumes of toxic, carcinogenic, and mutagenic wastewater, referred to as red water [5–8].

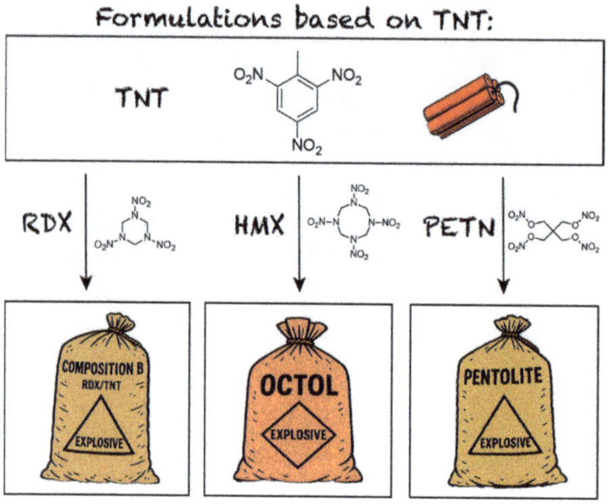

Figure 1. The overview of the most common formulations using TNT as the melt-castable component.

For large-scale TNT production, either an effective treatment method or a strategy to prevent red water formation is required [5]. However, treatment is costly and poses environmental risks, while process modifications to eliminate red water generation can compromise production efficiency [7]. A viable alternative is the development of new melt-castable energetic materials that avoid these issues altogether.

One of the primary criteria for a TNT replacement is thermal properties. A suitable substitute must exhibit a melting point between 80 and 120 °C, allowing the use of a melt-casting technique over a water steam bath. Moreover, its decomposition temperature should exceed the melting point by at least 100 °C, significantly reducing the risk of accidental ignition or decomposition during the melting process [9,10]. Another critical aspect is the energetic performance of the replacement material. The new compound should provide comparable or superior detonation velocity and pressure to maintain its effectiveness as an explosive. This ensures that the replacement not only matches but potentially exceeds the performance of TNT, making it viable for military and industrial applications. Safety is also a paramount consideration [9]. The alternative compound must be less sensitive to impact, friction, and shock, minimizing the likelihood of accidental detonation during handling, transportation, or storage [9]. Environmental and health concerns are also key drivers in the development of TNT alternatives. TNT is known for its persistent environmental contamination and toxicological impact on human health. A promising replacement should be non-toxic or exhibit significantly reduced toxicity and present a lower risk of contaminating soil and water ecosystems [11]. Compatibility with other explosives and additives is another fundamental requirement for TNT replacements. Since explosives are rarely used as pure substances in practical application, they are typically combined with other energetic materials (e.g., RDX, HMX) and functional additives such as binders, plasticizers, or metallic fuels. Compatibility testing is necessary to confirm that replacement material maintains its integrity and does not negatively impact the properties of the overall formulation [12]. Together, these criteria—thermal stability, energetic performance, safety, environmental sustainability, and compatibility—define the key parameters for the successful development of a new TNT replacement.

Nitropyrazoles are well suited as a scaffold for melt-castable explosives due to their favorable properties. Their high nitrogen content affords elevated heats of formation, whereas the four substitution sites on the pyrazole ring permit systematic property tuning [13,14]. A prominent example is 3,4-dinitropyrazole (DNP), which has been identified as a promising melt-castable matrix. DNP demonstrates a melting point (71 °C) [15,16] conducive to melting processes and offers improved detonation performance (8426 m s^{-1}) [15] compared to TNT. However, due to its higher viscosity and acidic NH function, the careful adjustment of the formulation is required to ensure suitable workability [17,18]. Our previous work has described the nitroalkyl and azidoalkyl derivatives of different nitropyrazoles as potential melt-castable explosives [19]. In this study, we want to introduce 3,5-dinitro-4-methylnitramino-1-methylpyrazole, patented by EURENCO [20], as a prospective TNT replacement. In addition to a comprehensive thermal and physicochemical analysis, we investigated its experimental heat of formation, compatibility with common energetic co-formulants, quantified its performance in the Small-Scale Reactivity Test (SSRT), and assessed its toxicity.

2. Results and Discussion

2.1. Synthesis

In the current synthesis protocol, the intermediate 4-chloro-3,5-dinitropyrazole (**1**), synthesized via pyrazole chlorination [21] followed by subsequent nitration [22,23], is converted into 3,5-dinitro-4-methylaminopyrazole monohydrate (**2**). This conversion is facilitated using an aqueous solution of methylamine within an optimized synthetic route [23,24]. Additionally, the anhydrous form of this compound can be obtained by desiccating the synthesized product at 50 °C overnight. This desiccation process results in a chromatic transition from yellow to orange. The methylamino group was subjected to nitration to produce 3,5-dinitro-4-methylnitraminopyrazole (**3**), following the procedure established by Dalinger [23]. The nitration was carried out using trifluoroacetic acid as the solvent and acetyl nitrate as the nitration agent, obtaining a high yield of 96%. Starting from **3**, a series of six ionic compounds (**3a–3f**) was formed through an acid–base reaction. To preserve the salts in the solid state, it was necessary to omit the use of water as a solvent. When water was used as a solvent for salt formation, it resulted in either the formation of a sticky mass or an extended time for solidification. In the concluding step, methylation was carried out with a yield of 69% by deprotonating **3** using hydrogen carbonate in water, followed by a reaction with dimethyl sulfate. The advantage here is the precipitation of 3,5-dinitro-4-methylnitramino-1-methylpyrazole (**4**) from the aqueous solution, enabling straightforward filtration. In cases where **4** fails to precipitate, extraction with ethyl acetate can be employed as an alternative method (Figure 2).

As an alternative approach, an effort was made to introduce the methyl group at an earlier stage (Figure 2). Compound **1** underwent successful methylation with dimethyl sulfate, resulting in 67% yield of 4-chloro-3,5-dinitro-1-methylpyrazole (**5**), which was further reacted with the aqueous methylamine solution, yielding 4-aminomethyl-3,5-dinitro-1-methylpyrazole (**6**) with 75% yield. Additionally, **6** can be obtained in a satisfactory yield of 80% through the methylation of compound **2** [25]. The final step, involving the nitration of the methylamino group to produce the targeted compound **4**, presented challenges within this alternative synthesis. The selected nitration conditions prevented the obtainment of compound **4** in a solid or powder form; instead, it was consistently obtained as a sticky mass. Therefore, it is advisable to reserve the methylation for the final step, as the resulting product precipitates from the aqueous neutral medium.

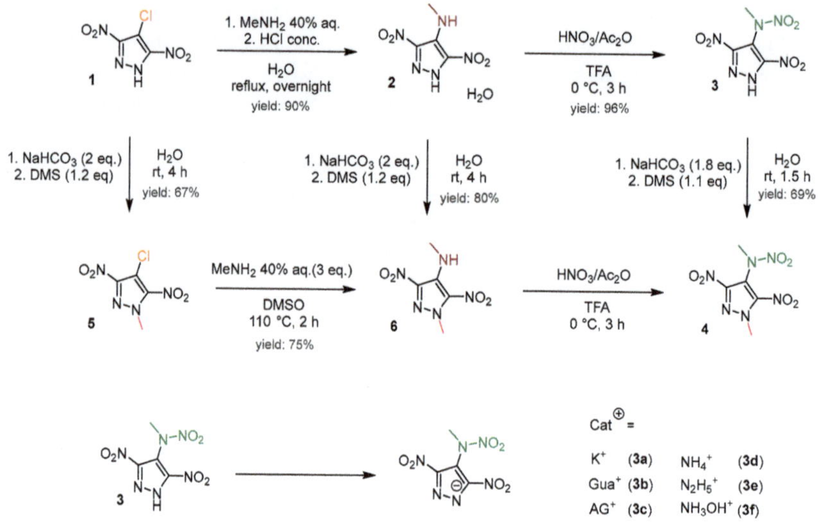

Figure 2. The synthesis overview of 3,5-dinitro-4-methylnitramino-1-methylpyrazole (**4**) and the salts of 3,5-dinitro-4-methylnitraminopyrzole (**3**).

2.2. Crystal Structures

Appropriate crystals for compounds **3**, **3a–3f** and **4–6** were obtained through recrystallization (**3**, **5**: ethanol; **3b**, **3d**, **3f**: water; **4**: acetonitrile; **6**: methanol) or directly from the reaction mixture (**3a**, **3c**, **3e**). The details of the X-ray measurements and refinements can be found in the Supporting Information. Additional information on the X-ray structure determinations has been deposited in the CCDC database with the following reference numbers: 2334166 (**3**), 2334159 (**3a**), 2334160 (**3b**), 2334157 (**3c**), 2334161 (**3d**), 2334158 (**3e**), 2334165 (**3f**), 2334164 (**4**), 2334163 (**5**), and 2334162 (**6**).

The ellipsoids, representing non-hydrogen atoms in all structures, are illustrated at the 50% probability level.

Figure 3 displays the crystal structures of the six salts of 3,5-dinitro-4-methylnitraminopyrazole (**3**). None of these compounds include water molecules in their structures. In general, the pyrazole anions in all salt crystal structures (**3a–3f**) presented here have nearly identical geometries. The N−N and C−N bond lengths within the pyrazole are all sustainably shorter than C−N single bonds (1.47 Å) but significantly longer than C=N double bonds (1.22 Å). This common feature explicitly demonstrates the delocalization of the negative charge over the aromatic ring system. Another important point to mention is that π-staggered arrangements have only been observed in the potassium and ammonium salts, where hydrogen bonding is of minor importance. In contrast, structures that are rich in hydrogen bonding, such the guanidinium salt **3b** or hydroxylammonium salt **3f**, are strongly determined by these intermolecular interactions.

The molecular unit cells and the extended structures of the two neutral compounds, **3** and **4**, are illustrated in Figure 4. Both pyrazoles crystallize in the monoclinic space groups *P*21 (**3**) and *P*21/c (**4**) with two and four molecules per unit cell, respectively. In both structures, it is observed that the nitro groups at the C1 and C3 atoms are only slightly twisted out of the plane of the pyrazole ring. However, the nitro groups of compound **4** exhibit a greater degree of twisting (NO_2−C1 17.28°, NO_2−C3 19.13°) compared to those of compound **3** (NO_2−C1 0.92°, NO_2−C3 10.17°) all in the same direction. This results in a different rotation of the methylnitramino group, which needs to be less twisted out of the pyrazole plane in compound **4** (53.27°) than in compound **3** (69.31°) due to the stronger twisting of the nitro groups. Additionally, the methyl group on the pyrazole ring of compound **4** protrudes slightly from the pyrazole plane (174.04–180°). The main difference between the two structures is that no intermolecular hydrogen bonds are formed in compound **4** compared to compound **3**. Compound **3** exhibits hydrogen bonds with a length of 2.196 Å between the proton on nitrogen N1 and the nitrogen N2 of the neighboring pyrazole, leading to the formation of regularly stacked layers. This observation likely contributes to the higher density observed in compound **3** (1.774 g/cm^3 @ 109 K) compared to compound **4** (1.696 g/cm^3 @ 102 K).

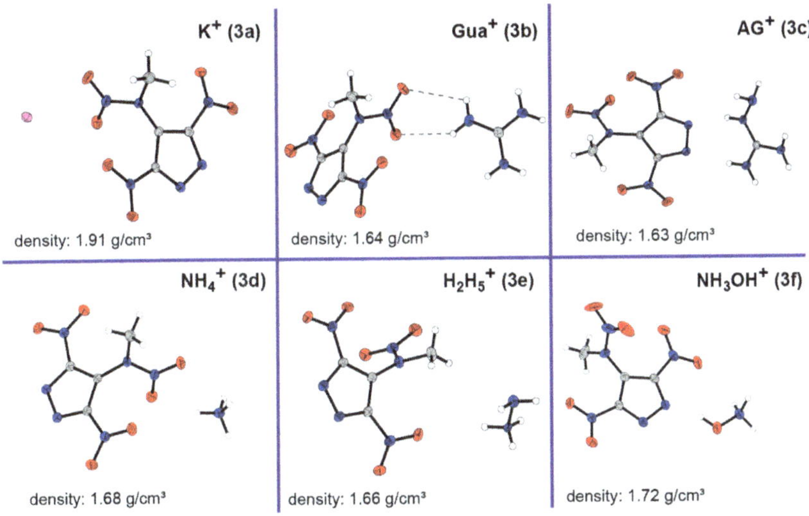

Figure 3. Molecular units of synthesized salts of compound **3**: **3a** potassium; **3b** guanidinium; **3c** aminoguanidinium; **3d** ammonium; **3e** hydrazinium; and **3f** hydroxylammonium.

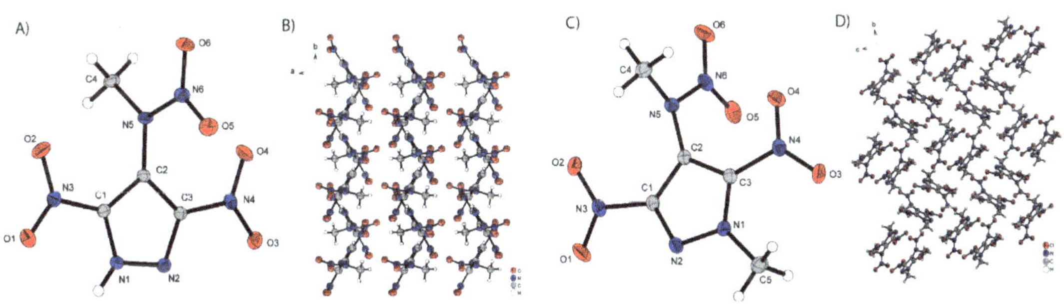

Figure 4. Molecular units of synthesized neutral compounds and its extended structure: (**A**) 3,5-dinitro-4-methylanitraminopyrazole (**3**), (**B**) extended structure of **3**, (**C**) 3,5-dinitro-methylnitramino-1-methylpyrazole (**4**) and (**D**) extended structure of **4**.

2.3. NMR Spectroscopy

NMR spectra, including ^{1}H, ^{13}C, and ^{14}N, were recorded for all compounds presented in this work. The comparison of the neutral compounds **2** and **3** reveals a downfield shift in the methyl group (**2**: ^{1}H 3.03 ppm, ^{13}C 33.4 ppm; **3**: ^{1}H 3.36 ppm, ^{13}C 40.5 ppm), attributed to the deshielding effects induced by the neighboring nitro group. Upon the further methylation of the pyrazole ring (compound **4**), the methyl group attached to the pyrazole undergoes stronger deshielding (^{1}H 4.35 ppm, ^{13}C 43.2 ppm) compared to the methyl group attached to the nitramine (^{1}H 3.35 ppm, ^{13}C 40.2 ppm). The methyl group on the nitramine of compound **4** is slightly more shielded compared to **3**, resulting in its signal overlap with signals from DMSO-*d6* when recording the ^{13}C spectrum in DMSO-*d6* at low concentrations. To clearly distinguish all methyl groups in the ^{13}C spectrum, even at lower concentrations, the NMR spectrum can also be recorded in acetone-*d6*. The greater shift of the methyl group on the pyrazole ring compared to that on the nitramine is further confirmed by compounds **5** (^{1}H 4.28 ppm, ^{13}C 43.5 ppm) and **6** (^{1}H 4.16 ppm, ^{13}C 43.0 ppm).

Additionally, for compound **4**, ^{15}N and ^{1}H ^{15}N HMBC NMR spectra were obtained. The ^{15}N ^{1}H-decoupled spectra of compound 4 reveal only five signals instead of the expected six. To locate the missing signal, the ^{1}H ^{15}N-coupled spectrum is employed (Figure 5). Focusing on the signal at −33 ppm reveals the presence of two nitro groups. The nitro group N6 induces a quartet splitting pattern by coupling with the methyl group. A singlet appears precisely at the center of the quartet, indicative of the nitro group N4. Additionally, the third nitro group N5 appears as a singlet at −27 ppm. The remaining three singlets each exhibit quartet splitting. The observed splitting in the pyrazole signals is attributed to the coupling with the methyl group on the pyrazole ring, while that of the N3 signal stems from the coupling with the methyl group on the nitramine. Signal assignment was accomplished utilizing the ^{1}H ^{15}N HMBC spectrum.

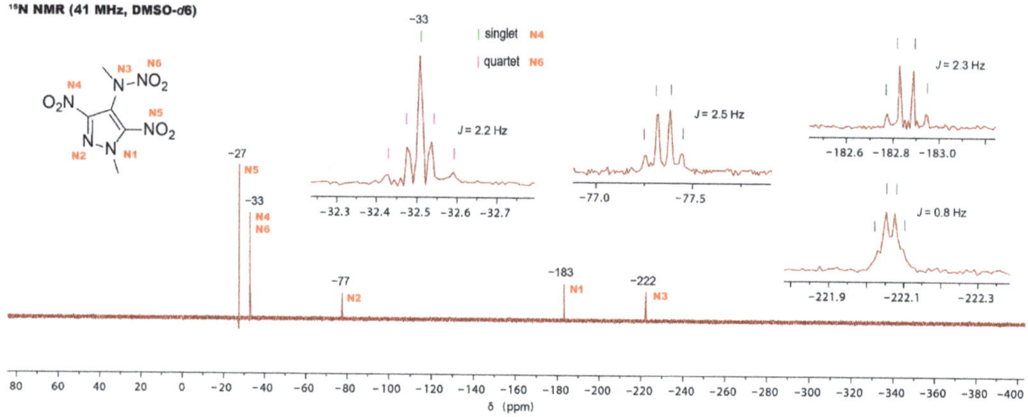

Figure 5. The proton-coupled ^{15}N NMR spectrum of compound **4**, showing the splitting of the signals.

2.4. Physiochemical Properties

The physicochemical properties, including thermal behavior, sensitivities to impact, friction, and electrostatic discharge, as well as the detonation parameters of the six salts of compound 3 and neutral compounds 3, **4**, and **6**, are summarized in Table 1 (more detailed tables can be found in the SI, Tables S8 and S9).

A comparison of the salts with each other and with the neutral compound **3** reveals different trends. The potassium salt (**3a**) has the highest thermal stability (217 °C) and density at room temperature (1.906 g/cm^3), but has the lowest detonation velocity (7296 m/s). Conversely, the hydroxylammonium salt (**3f**) has the highest detonation velocity of 8470 m/s and a commendable density of 1.724 g/cm^3, albeit with the lowest decomposition point at 147 °C. Only the guanidinium (**3b**) and aminoguanidinium (**3c**) salts show an endothermic event just before decomposition. In terms of decomposition temperatures, all salts, with the exception of potassium (217 °C), show a similar range (147–169 °C) compared to the neutral compound **3** (157 °C). However, the formation of salts (excluding potassium and hydroxylammonium salts) leads to a decrease in density compared to the neutral compound.

Regarding sensitivities, the guanidinium salt (**3b**) is characterized by an insensitivity towards impact and friction (confirmed twice with different batches). Conversely, the other salts show an increased sensitivity to impact compared to the neutral compound **3**. While only the potassium (**3a**) and hydrazinium (**3e**) salts show a higher sensitivity to friction compared to the neutral compound, the others show a lower sensitivity.

Table 1. Physiochemical properties and detonation parameters of all synthesized energetic compounds compared to TNT and DNAN.

	IS [a] [J]	*FS* [b] [N]	T_{endo} [c] [°C]	T_{exo} [d] [°C]	*ρ* [e] [g cm^{-3}]	$\Delta_f H°$ [f] [kJ·mol^{-1}]	$D_{C\text{-}J}$ [g] [m·s^{-1}]	$p_{C\text{-}J}$ [h] [GPa]
3a	6	80	/	217	1.906	−446.5	7296	21.7
3b	>40	>360	140	158	1.638	60.0	7719	22.7
3c	8	168	116	148	1.628	165.7	7848	23.3
3d	3	192	/	169	1.681	101.9	8129	26.8
3e	5	120	/	165	1.660	248.5	8261	27.3
3f	5	192	/	147	1.724	157.7	8470	30.5
3	8	144	/	157	1.725	145	8228	28.8
6	>40	>360	133	203	1.629	76.1	7273	19.6
4	15	>360	77	190	1.648	135.2/106.4 [j]	7721/7682 [j]	24.1/23.8 [j]
TNT [i]	**15**	**>360**	**81**	**289**	**1.65**	**−185**	**6950**	**20.5**
DNAN [26]	**>40**	**>360**	**94**	**315**	**1.59**	**−177**	**6705**	**16.1**

[a] Impact sensitivity (BAM drop hammer, method 1 of 6); [b] friction sensitivity (BAM friction tester, method 1 of 6); [c] endothermic event (DTA, β = 5 °C·min^{-1}); [d] temperature of decomposition (DTA, β = 5 °C·min^{-1}); [e] density at 298 K recalculated from X-ray data; [f] heat of formation (calculated using the atomization method and CBS-4M enthalpies); [g] detonation velocity; [h] detonation pressure; [i] determined at LMU; [j] determined experimentally (bomb calorimetry).

Among the neutral compounds, compound **3** is characterized by the highest density (1.725 g/cm^3) and detonation velocity (8228 m/s). However, it has no melting point and exhibits a relatively low decomposition temperature of 157 °C. It also shows increased sensitivity to impact (8 J) and friction (144 N). The methylation of compound **2** on the pyrazole ring yields compound **6**, a melt-castable derivative with an acceptable decomposition temperature of 203 °C and insensitivity to mechanical stimuli. Nevertheless, its melting point is somewhat too high for practical applications (133 °C) and its detonation velocity of 7273 m/s is rather low. The combination of compounds **2** and **3** to form compound **4** leads to a promising TNT replacement. Compound **4** has a sensitivity (IS: 15 J, FS: >360 N) and a density (1.648 g/cm^3) comparable to TNT and a suitable melting point of 77 °C (Figure 6). Although its decomposition temperature (190 °C) is lower than that of TNT (289 °C) or DNAN (315 °C), the necessary safety margin of at least 100 °C between the melting and decomposition point is maintained. It is significant that compound **4** has a detonation velocity of 7721 m/s and a detonation pressure of 24.1 GPa, both of which are higher than those of TNT or DNAN. Since the heat of formation influences the detonation parameters [1], it was additionally determined experimentally for compound **4** using a bomb calorimetry (details on the measurement can be found in SI, S4 Bomb Calorimeter). A comparison between the experimental (106.4 kJ/mol) and calculated values (135.2 kJ/mol) reveals that the calculation yielded a slightly more endothermic value (Table 1). The recalculation of the detonation parameters using the experimental heat of formation therefore resulted in slightly lower detonation velocity (7682 m/s) and pressure (23.8 GPa).

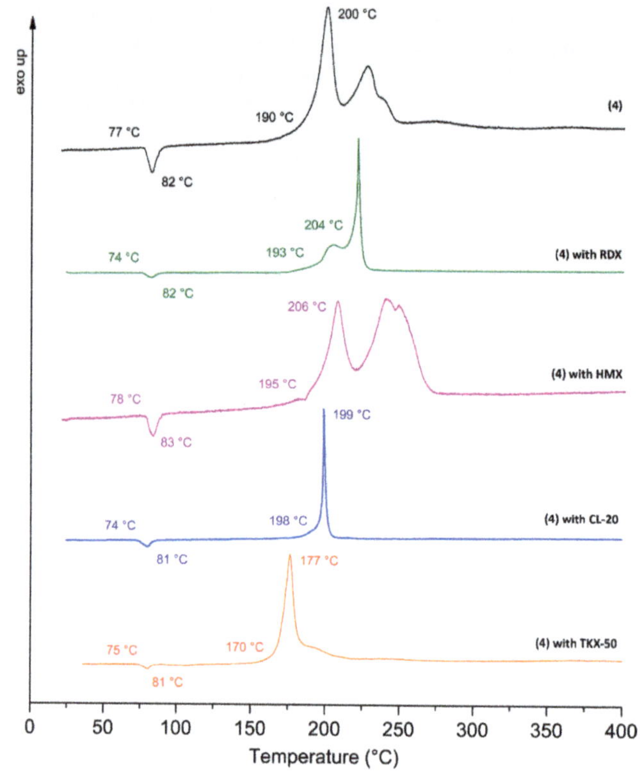

Figure 6. DTA compatibility measurements of **4** with RDX, HMX, CL-20, and TKX-50.

Compound **4**, recognized as a potential alternative to TNT, underwent compatibility testing with RDX, HMX, CL-20, and TKX-50 (Figure 6). Compatibility measurements were performed using differential thermal analysis (DTA). The results demonstrate a high degree of compatibility between compound **4** and both RDX and CL-20, characterized by a minimal difference of 0 to 4 °C between their melting points and decomposition temperatures. HMX and TKX-50 demonstrate a deviation of 1 °C only at the melting point. However, compatibility for HMX at the decomposition temperature is rated moderate (6 °C difference), albeit close to the lower limit of good compatibility. TKX-50 demonstrates incompatibility at the decomposition point with compound **4**, as indicated by a temperature difference of 23 °C (for additional information, see SI, S8 Compatibilities).

2.5. Small-Scale Shock Reactivity Test

The SSRT test evaluates the explosiveness of compounds on a small scale of ~500 mg and a fixed volume of 283 mm^3. An explosive compound is filled into a steel cylinder with a borehole and pressed on top of an aluminum witness block. The compound is filled into the hole and pressed with a weight of 3 t. A commercial detonator is used to initiate the explosion, creating a dent in the aluminum block. The dent volume can be measured using optical topography and can then be compared to other energetic materials [27]. Several factors such as overall energy release, detonation velocity, and pressure are combined in this test setup to assess the compound's general explosiveness. The tested compound **4** significantly outperforms TNT by 22% in dent volume, although the density of compound **4** is the same (Table 2). The values are still comparable to PETN, which has a higher density than compound **4** (for addition information see SI, S7 SSRT).

Table 2. Results of SSRT for **4** compared to PETN and TNT.

	4	PETN	TNT
Dent−Volume [mm^3]	1037.34	1107.81	845.89
m [g] [a]	443	478	443
ρ [g cm^{-3}] [b]	1.648	1.778	1.648
$p_{C\text{-}J}$ [GPa] [c]	24.1 [e]	30.8 [e]	19.4 [e]
V_{det} [m s^{-1}] [d]	7721 [e]	8429 [e]	6839 [e]

[a] Used mass; [b] density; [c] detonation pressure; [d] detonation velocity; [e] calculated with EXPLO5_V6.05.

2.6. *Ames Test*

Performing an Ames test on a new explosive is essential to identify potential risks arising from exposure to that substance. Given the wide range of applications for explosives in the military and civilian sectors, it is imperative to ensure if genetic mutations and associated health risks to both humans and the environment occur. Carrying out an Ames test enables the early detection of potential hazards and facilitates the implementation of appropriate measures to ensure human and environmental safety. The test uses specific bacterial strains with genetic mutations that prevent them from producing certain amino acids. When a substance with mutagenic potential is tested, it can reverse these mutations, leading to bacterial growth. The mutagenic activity of the tested substance can be concluded by observing the extent of bacterial growth (for more information, see SI, S9 Ames Test).

Compound **4** showed a mutagenic effect in all strains in both the assay without and with metabolic activation. The increased number of revertant colonies in the assay without metabolic activation was observed at the concentrations of 4.11–0.46 µg/plate for TA98 and TA1537 strains, 1.37–0.46 µg/plate for TA100 and TA1535, and 37.03–0.46 µg/plate for *E. coli*. After metabolic activation, the number of revertant colonies increased for TA98 at concentrations ranging from 12.3 to 0.4 µg/plate, for TA100 from 4.11 to 0.46 µg/plate, for **TA1535** from 4.11 to 1.37 µg/plate, for TA1537 from 12.3 to 1.37 µg/plate, and for *E. coli* from 37.03 to 4.11 µg/plate.

3. Materials and Methods

All used materials and methods and the detailed experimental part can be found in Supplementary Materials.

4. Conclusions

In this study, six salts of 3,5-dinitro-4-methylnitraminopyrazole (**3**) were synthesized and characterized. The potassium salt showed a remarkable thermal stability, with a decomposition temperature of 217 °C. The hydroxylammonium salt (**3f**) displayed the highest detonation velocity of 8470 m/s. The guanidinium salt (**3b**) showed the desirable complete insensitivity to mechanical stimuli. The aminoguanidinium (**3c**), ammonium (**3d**) and hydrazinium (**3e**) salts showed properties very similar to those of the neutral compound (**3**). The most interesting new compound is the melt-castable energetic material 3,5-dinitro-4-methylnitramino-1-methylpyrazole (**4**). The substitution of acidic protons with methyl groups weakens intermolecular interactions just enough to result in a lowered melting point of 77 °C while still retaining sufficient thermal decomposition temperature of 190 °C. The introduction of the methyl group to the pyrazole ring in the final synthesis step yields the most optimal results, as the product readily precipitates from water. Precautions should be taken due to the observation of potentially mutagenic effects in the Ames test. Compound **4** demonstrates compatibility with RDX, HMX, and CL-20, indicating its potential application as a drop-in TNT replacement. The sensitivity to mechanical stimuli and density at room temperature of compound **4** are comparable to those of TNT, while surpassing TNT's detonation velocity by more than 10%.

Supplementary Materials: The following supporting information can be downloaded at: https://www.mdpi.com/article/10.3390/molecules30132796/s1, Schemes S1–S3: Synthesis of the presented compounds; Tables S1–S4: Crystallographic data and structure refinement details of the prepared compounds; Table S5: CBS-4M electronic enthalpies for atoms C, H, N, and O and their literature values; Table S6: CBS-4M results and calculated gas-phase enthalpies; Table S7: Heat of formation calculation results for compounds **3**, **4**, and **6**; Tables S8 and S9: Physicochemical properties and detonation parameter of prepared energetic compounds; Table S10: Report of the experimental data from bomb calorimeter from the Combustion Calorimeter Tool—Experiment Mode (NIST); Table S11: Enthalpies of formation of the reaction products; Table S12: Thermodynamic characteristics of (**4**), including combustion enthalpies, change in gas mole, and formation enthalpy and energy; Tables S13 and S14: SSRT results of **4**; Table S15: Criteria for compatibilities; Table S16: Compatibilities of compounds **4** with RDX, HMX, and CL-20; Figures S1–S10: X-ray structures of prepared compounds; Figures S11 and S12: ^{15}N NMR spectra of **4**; Figures S13–S17: DTA curves; Figure S18: Compound **4** during the melting process; Figure S19: Setup of the SSRT experiment; Figures S20–S23: DTA curves of compatibility measurements. References [23,28–53] are cited in the Supplementary.

Author Contributions: Methodology, E.R. and J.S.; investigation, E.R.; practical work, A.H.S., H.R.W. and L.B.; writing—original draft, E.R.; resources, A.D.; supervision, T.M.K. All authors have read and agreed to the published version of the manuscript.

Funding: The financial support of this work by the Ludwig-Maximilian University (LMU), EMTO GmbH, BIAZZI SA, and Eurenco in France is greatly acknowledged.

Institutional Review Board Statement: Not applicable.

Informed Consent Statement: Not applicable.

Data Availability Statement: The data presented in this study are available in the article and Supplementary Materials.

Acknowledgments: The authors thank Stefan Huber for his help with the sensitivity measurements, Karaghiosoff for ^{15}N NMR measurements, and Marcus Lommel for single-crystal measurements and general support.

Conflicts of Interest: T.M. Klapötke, E. Reinhardt and J. Stierstorfer work part-time at EMTO GmbH. A. Delage is an employee of Eurenco.

References

1. Klapoetke, T.M. *Chemistry of High-Energy Materials*, 6th ed.; De Gruyter: Berlin, Germany; Boston, MA, USA, 2022.
2. Song, S.; Chen, F.; Wang, Y.; Wang, K.; Yan, M.; Zhang, Q. Accelerating the Discovery of Energetic Melt-Castable Materials by a High-Throughput Virtual Screening and Experimental Approach. *J. Mater. Chem. A* **2021**, *9*, 21723–21731. [CrossRef]
3. Agrawal, J.P. *High Energy Materials*; Wiley: Weinheim, Germany, 2010.
4. Urbański, T.; Laverton, S. *Chemistry and Technology of Explosives*; Macmillan: New York, NY, USA, 1964.
5. Millar, R.W.; Anthony, A.W.; Hamid, J.; Endsor, R.M. *Elimination of Redwater Formation from TNT Manufacture*; QinetiQ: Farnborough, UK; Arlington, VA, USA, 2007.
6. Freeman, D.; Colitti, O. Removal of Explosives from Load-Assemble-Pack Wastewater (Pink Water) Using Surfactant Technology. In Proceedings of the 36th Industrial Waste Conference, Purdue University, West Lafayette, IN, USA, 12–14 May 1981.
7. Szopińska, M.; Prasuła, P.; Baran, P.; Kaczmarzyk, I.; Pierpaoli, M.; Nawała, J.; Szala, M.; Fudala-Książek, S.; Kamieńska-Duda, A.; Dettlaff, A. Efficient Removal of 2,4,6-Trinitrotoluene (TNT) from Industrial/Military Wastewater Using Anodic Oxidation on Boron-Doped Diamond Electrodes. *Sci. Rep.* **2024**, *14*, 4802. [CrossRef] [PubMed]
8. Gilbert, E.E.; Siele, V.I.; Chandler, C.D.; Mundy, R.A.; Spencer, J.R.; Gibson, G.R.; Bolleter, W.T. New Procedures for Purifying TNT. *Propellants Explo. Pyrotec.* **1982**, *7*, 150–154. [CrossRef]

9. Larin, A.A.; Bystrov, D.M.; Fershtat, L.L.; Konnov, A.A.; Makhova, N.N.; Monogarov, K.A.; Meerov, D.B.; Melnikov, I.N.; Pivkina, A.N.; Kiselev, V.G.; et al. Nitro-, Cyano-, and Methylfuroxans, and Their Bis-Derivatives: From Green Primary to Melt-Cast Explosives. *Molecules* **2020**, *25*, 5836. [CrossRef] [PubMed]
10. Ravi, P.; Badgujar, D.M.; Gore, G.M.; Tewari, S.P.; Sikder, A.K. Review on Melt Cast Explosives. *Propellants Explo. Pyrotec.* **2011**, *36*, 393–403. [CrossRef]
11. Benz, M.; Delage, A.; Klapötke, T.M.; Kofen, M.; Stierstorfer, J. A Rocky Road toward a Suitable TNT Replacement—A Closer Look at Three Promising Azoles. *Propellants Explo. Pyrotec.* **2023**, *48*, e202300042. [CrossRef]
12. Li, J.; Chen, J.; Hwang, C.; Lu, K.; Yeh, T. Study on Thermal Characteristics of TNT Based Melt-Cast Explosives. *Propellants Explo. Pyrotec.* **2019**, *44*, 1270–1281. [CrossRef]
13. Zhang, S.; Gao, Z.; Lan, D.; Jia, Q.; Liu, N.; Zhang, J.; Kou, K. Recent Advances in Synthesis and Properties of Nitrated-Pyrazoles Based Energetic Compounds. *Molecules* **2020**, *25*, 3475. [CrossRef]
14. Xie, H.; Zhang, X.; Jiang, T.; Zhu, Y.; Zhou, L. Rheological Behavior of DNP/HMX Melt-Cast Explosives with Bimodal and Trimodal Particle-Size Distributions. *Polymers* **2023**, *15*, 1446. [CrossRef]
15. Bölter, M.F.; Harter, A.; Klapötke, T.M.; Stierstorfer, J. Isomers of Dinitropyrazoles: Synthesis, Comparison and Tuning of Their Physicochemical Properties. *ChemPlusChem* **2018**, *83*, 804–811. [CrossRef]
16. Kuehl, V.A.; Cleveland, A.H.; Snyder, C.J.; Chavez, D.E. Synthesis and Energetic Properties of N-Substituted 3,4- and 3,5-Dinitropyrazoles. *ACS Omega* **2023**, *8*, 18408–18413. [CrossRef] [PubMed]
17. Yang, C.; Yang, Y.; Zhu, S.; Zhang, S.; Liu, Y.; Chen, Y. Anisotropic Shock Response of 3,4-Dinitropyrazole Revealed by First-Principles Calculations. *J. Phys. Chem. A* **2025**, *129*, 695–704. [CrossRef] [PubMed]
18. Tang, Y.; He, C.; Gao, H.; Shreeve, J.M. Energized Nitro-Substituted Azoles through Ether Bridges. *J. Mater. Chem. A* **2015**, *3*, 15576–15582. [CrossRef]
19. Reinhardt, E.; Lenz, T.; Bauer, L.; Stierstorfer, J.; Klapötke, T.M. Synthesis and Characterization of Azido- and Nitratoalkyl Nitropyrazoles as Potential Melt-Cast Explosives. *Molecules* **2023**, *28*, 6489. [CrossRef]
20. Klapötke, T.M.; Stierstorfer, J.; Reinhardt, E. Nitramine Pyrazole Energetic Compounds. WO 2023/072812 A1 Patent, 14 December 2023.
21. Baus, U.; Reuther, W. Verfahren Zur Herstellung von 4-Chloropyrazolen 1990. DE 3840342 A1 Patent, 23 June 2021.
22. Fischer, D.; Gottfried, J.L.; Klapötke, T.M.; Karaghiosoff, K.; Stierstorfer, J.; Witkowski, T.G. Synthesis and Investigation of Advanced Energetic Materials Based on Bispyrazolylmethanes. *Angew. Chem. Int. Ed.* **2016**, *55*, 16132–16135. [CrossRef]
23. Dalinger, I.L.; Vatsadze, I.A.; Shkineva, T.K.; Popova, G.P.; Shevelev, S.A. Efficient Procedure for High-Yield Synthesis of 4-Substituted 3,5-Dinitropyra- Zoles Using 4-Chloro-3,5-Dinitropyrazole. *Synthesis* **2012**, *44*, 2058–2064. [CrossRef]
24. Reinhardt, E.; Thamm, S.; Stierstorfer, J.; Klapötke, T.M. Alkyl-Bridged Nitropyrazoles—Adjustment of Performance and Sensitivity Parameters. *Eur. J. Org. Chem.* **2023**, *26*, e202300304. [CrossRef]
25. Dalinger, I.L.; Vatsadze, I.A.; Shkineva, T.K.; Popova, G.P.; Shevelev, S.A. The Specific Reactivity of 3,4,5-Trinitro-1H-Pyrazole. *Mendeleev Commun.* **2010**, *20*, 253–254. [CrossRef]
26. Klapötke, T.M.; Penger, A.; Pflüger, C.; Stierstorfer, J. Melt-Cast Materials: Combining the Advantages of Highly Nitrated Azoles and Open-Chain Nitramines. *New J. Chem.* **2016**, *40*, 6059–6069. [CrossRef]
27. Bauer, L.; Benz, M.; Klapötke, T.M. Linear Correlation Between Confined Explosive Quantity and Dent Volume of an Underlying Aluminium Block Using the SSRT Setup. *Propellants Explo. Pyrotec.* **2022**, *47*, e202100332. [CrossRef]
28. *CrysAlisPro Software*, Rigaku Corporation: Wroclaw, Poland, 2009. Available online: http://www.rigaku.com (accessed on 25 May 2025).
29. Sheldrick, G.M. SHELXT Integrated space-group and structure determination. *Acta Cryst.* **2015**, *A71*, 3–8. [CrossRef] [PubMed]
30. Dolomanov, O.V.; Bourhis, L.J.; Gildea, R.J.; Howard, J.A.K.; Puschmann, H.J. OLEX2: A Complete Structure Solution, Refinement and Analysis Programm. *Appl. Cryst.* **2009**, *42*, 339–341. [CrossRef]
31. *SCALE3 ABSPACK—An Oxford Diffraction Program*, Version 1.0.4; Rigaku Corporation: Wroclaw, Poland, 2009. Available online: http://www.rigaku.com (accessed on 25 May 2025).
32. *APEX3*, Version 2018.7-2; Bruker AXS Inc.: Madison, WI, USA, 2018. Available online: http://www.bruker.com (accessed on 25 May 2025).
33. Frisch, M.J.; Trucks, G.W.; Schlegel, H.B.; Scuseria, G.E.; Robb, M.A.; Cheeseman, J.R.; Scalmani, G.; Barone, V.; Petersson, G.A.; Nakatsuji, H.; et al. *Gaussian 16*, Revision A.04; Gaussian, Inc.: Wallingford, CT, USA, 2016.
34. Montgomery, J.A.; Frisch, M.J.; Ochterski, J.W.; Petersson, G.A. A complete basis set model chemistry. VII. Use of the minimum population localization method. *J. Chem. Phys.* **2000**, *12*, 6532–6542.
35. Curtiss, L.A.; Raghavachari, K.; Redfern, P.C.; Pople, J.A. Assessment of Gaussian-2 and density functional theories for the computation of enthalpies of formation. *J. Chem. Phys.* **1997**, *106*, 1063–1079. [CrossRef]
36. Rice, B.M.; Pai, S.V.; Hare, J. Predicting heats of formation of energetic materials using quantum mechanical calculations. *Combust. Flame* **1999**, *118*, 445–458. [CrossRef]

37. Byrd, E.F.C.; Rice, B.M. Improved Prediction of Heats of Formation of Energetic Materials Using Quantum Mechanical Calculations. *J. Phys. Chem. A* **2006**, *110*, 1005–1013. [CrossRef] [PubMed]
38. Linstrom, J.; Mallard, W.G. National Institute of Standards and Technology Standard Reference Database Number 69. Available online: http://www.webbook.nist.gov/chemistry (accessed on 25 May 2025).
39. Trouton, F. IV. On molecular latent heat. *Philosophical Magazine Series 5* **1884**, *18*, 54–57. [CrossRef]
40. Westwell, M.S.; Searle, M.S.; Wales, D.J.; Williams, D.H. Empirical Correlations between Thermodynamic Properties and Intermolecular Forces. *J. Am. Chem. Soc.* **1995**, *117*, 5013–5015. [CrossRef]
41. Rossini, F.D. *Experimental Thermochemistry: Measurement of Heats of Reaction*; Interscience Publishers Inc.: New York, NY, USA, 1956.
42. Paulechka, E.; Riccardi, D. *Combustion Calorimetry Tool—NIST Standard Reference Database 206*; National Institute of Standard and Technology: Gaithersburg, MD, USA, 2019.
43. Koehler, J.; Meyer, R.; Homburg, A. *Explosivstoffe*, 6th ed.; Wiley: Weinheim, Germany, 2007.
44. *STANAG 4147*; Chemical Compatibility of Ammunition Components with Explosives (Non-Nuclear Application). NATO: Brussels, Belgium, 2001.
45. Bienta. Available online: https://bienta.net (accessed on 25 May 2025).
46. Bundesanstalt für Materialforschung und -Prüfung. Available online: http://www.bam.de (accessed on 25 May 2025).
47. *STANAG 4489*; Explosives, Impact Sensitivity Tests. NATO: Brussels, Belgium, 1999.
48. *Standardarbeitsanweisung 4-5.1.02*; Ermittlung der Explosionsgefährlichkeit, hier der Schlagempfindlichkeit mit dem Fallhammer. WIWEB: Erding, Germany, 2002.
49. *STANAG 4487*; Explosives, Friction Sensitivity Tests. NATO: Brussels, Belgium, 2002.
50. *Standardarbeitsanweisung 4-5.1.03*; Ermittlung der Explosionsgefährlichkeit oder der Reibeempfindlichkeit mit dem Reibeapparat. WIWEB: Erding, Germany, 2002.
51. OZM Research. Available online: http://www.ozm.cz (accessed on 25 May 2025).
52. Impact: Insensitive > 40 J, Less Sensitive ≤35 J, Sensitive ≤4 J, Very Sensitive ≤3 J; Friction: Insensitive > 360 N, Less Sensitive = 360 N, Sensitive 80 N, Very Sensitive ≤80 N, Extreme Sensitive ≤10 N; According to the UN Recommendations on the Transport of Dangerous Goods (+) Indicates: Not Safe for Transport. Available online: https://www.un-ilibrary.org/content/books/9789210019057 (accessed on 25 May 2025).
53. Dalinger, I.L.; Vatsadze, I.A.; Shkineva, T.K.; Papova, G.P.; Shevelev, S.A.; Nelyubina, Y.V. Synthesis and Comparison of the Reactivity of 3,4,5-1*H*-Trinitropyrazole and Its *N*-Methyl Derivative. *J. Heterocycl. Chem.* **2013**, *50*, 911–924. [CrossRef]

molecules

MDPI

Article

Synthesis and Characterization of 1-Hydroxy-5-Methyltetrazole and Its Energetic Salts

Lukas J. Eberhardt, Maximilian Benz, Jörg Stierstorfer and Thomas M. Klapötke *

Department of Chemistry, Ludwig-Maximilian University of Munich, 81377 Munich, Germany; lukebch@cup.uni-muenchen.de (L.J.E.); jstch@cup.uni-muenchen.de (J.S.)
* Correspondence: tmk@cup.uni-muenchen.de

Abstract: The objective of this work was the synthesis and characterization of novel, insensitive high explosives. 1-hydroxy-5-methyltetrazole served as both a scaffold and anion for preparing various nitrogen-rich energetic salts. The compounds were characterized using ^{1}H and ^{13}C NMR spectroscopy, high-resolution mass spectrometry, elemental analysis, low-temperature single-crystal X-ray diffraction, and IR spectroscopy. Thermal stability was investigated via differential thermal analysis (DTA). Sensitivities towards mechanical stimuli were measured using a BAM drop hammer for impact sensitivity and a BAM friction apparatus for friction sensitivity, employing one of six testing procedures. Energetic performance parameters were calculated using the EXPLO5 code, incorporating room-temperature X-ray densities and solid-state heats of formation obtained via CBS-4M calculations using the Gaussian 16 program.

Keywords: tetrazole; energetic materials; cycloaddition; oxidation

1. Introduction

The majority of secondary explosives exhibit significant shortcomings. 1,3,5-Trinitrotoluene (TNT) is a melt-castable explosive, with a melting point of 80 °C. It is possible to melt TNT in a steam bath and cast it into its desired shapes; in doing so, an ideal filling is obtained. To be considered a melt-castable explosive, the melting temperature should be in the range of 75–115 °C, while decomposition temperatures should be higher than 180 °C [1,2]. TNT has been produced and used in explosive charges and formulations for over a century. Despite the affordability, accessibility, high decomposition temperatures, and low sensitivity towards mechanical stimuli, TNT suffers from two problems and therefore needs to be substituted. Firstly, TNT is toxic and a potential human carcinogen. It was shown that the groundwater and soil near production sites or training grounds are contaminated with TNT, by-products like dinitrotoluenes and degradation products, posing a threat to the environment [3–5]. The second problem is the weak detonation performance compared to other military explosives like RDX (1,3,5-trinitro-1,3,5-triazinane) or HMX (1,3,5,7-tetranitro-1,3,5,7-tetrazocane). RDX and HMX—while having a high detonation velocity—are toxic and pose an environmental threat on their own and through their degradation products [5,6]. Current research trends in energetic materials focus on developing more environmentally friendly explosives that are less toxic while maintaining or enhancing detonation performance and sensitivity. Another important consideration is the 'green' synthesis of new energetic materials, which should ideally have a high atom economy and use readily available starting materials. Furthermore, hazardous reagents and solvents should be avoided in the synthesis to ensure safer, more sustainable processes [7].

 https://doi.org/10.3390/molecules30132766

Due to their favorable energetic properties, tetrazoles have attracted considerable interest as scaffolds for high-energy-density materials (HEDMs). Among the azoles, tetrazoles exhibit the second highest heat of formation after pentazoles, resulting in good energetic performance. Their high nitrogen content ensures that the primary decomposition product is the non-toxic elemental dinitrogen [1,8]. 5-substituted tetrazoles are synthetically accessible via various protocols, including the 1,3-dipolar cycloaddition of nitriles with azides [9–12] or ring-closing reactions of azidoimines [13–16]. Introducing N-hydroxy functionalities into tetrazoles can significantly enhance the physicochemical properties of neutral hydroxytetrazoles and their salts compared to the parent compound. This is evidenced by an increase in oxygen balance and density and higher overall detonation velocities [17–20]. Hydroxytetrazoles can be synthesized primarily via two synthetic approaches. The direct oxidation of tetrazoles was first reported by Begtrup et al. in 1995; using sodium perborate in pivalic acid at 100 °C, 5H-tetrazole was converted to 1-hydroxytetrazole and 2-hydroxytetrazole in a 2:1 molar ratio and a combined yield of 50% [21]. In 1999, Giles et al. pioneered the use of Oxone® ($KHSO_5 \cdot KHSO_4 \cdot K_2SO_4$), buffered at pH 7.5, for the direct oxidation of tetrazoles. In this case, ethyl tetrazole-5-carboxylate was selectively converted to ethyl-2-hydroxytetrazole-5-carboxylate [22]. The utilization of Oxone® has now become standard methodology for the introduction of N-hydroxy functionalities in tetrazole rings. The oxidation of 5-substituted tetrazoles results in mixtures of regioisomers if the 5-substituent is an electron-donating group (EDG) [21]. However, if the 5-position bears an electron-withdrawing group (EWG), such as N_3 [19], CO_2Et [22], or NO_2 [23], the reaction proceeds regioselectively towards N2. In addition to direct oxidation, 1-hydroxy-tetrazoles can be synthesized utilizing the azidooxime 1-hydroxytetrazole tautomerism. Azidooximes are readily available starting from aldehydes or nitriles. First, the aldehyde is converted to the corresponding oxime by reacting with hydroxylamine, followed by chlorination to form an oxime chloride. The cascade for nitriles is similar: the nitril reacts with hydroxylamine, forming an amidoxime, which can be diazotized with sodium nitrite in hydrochloric acid, yielding the oxime chloride. The subsequent reaction of the oxime chlorides with NaN_3 results in the formation of the azidooxime, which is cyclized under acidic conditions, yielding 1-hydroxy-tetrazoles [20,24–27]. The synthesis of 1-hydroxy-5-methyl tetrazole was first documented by Bettinetti et al. in 1956, via the reaction of hydrazoic acid and ethyl nitrolic acid. In addition to the neutral compound, the silver salt was also prepared. However, neither compound was characterized completely, nor were their energetic parameters explored [28]. For safety reasons, in this study, 1-hydroxy-5-methyltetrazole was synthesized by the oxidation of 5-methyltetrazole.

2. Results and Discussion

Parts of this work were already published at the NTREM conference [29].

2.1. Synthesis

5-Methyltetrazole was first synthesized by Oberhummer in 1933 via the diazotization of acetimidohydrazide using ethyl nitrate [15]. The first reported synthesis of 5-methyltetrazole via 1,3-dipolar cycloaddition was carried out by Mihina et al. in 1950, achieving a yield of 76% from the reaction of acetonitrile with hydrazoic acid [30]. More recent methods use the reaction of sodium azide with acetonitrile in the presence of catalysts, such as $NiFe_2O_4$, Cu(II)-NaY zeolite, or CuO/aluminosilicate, in DMF, yielding up to 99% of the desired product [31–33].

In 2001, Sharpless et al. significantly improved the preparation of 5-substituted tetrazoles by performing the 1,3-dipolar cycloaddition of organic nitriles with sodium azide in water, using zinc Lewis acids as catalysts, a more environmentally friendly alternative to conventional protocols, which typically employ toxic organic solvents, such as DMF (Scheme 1) [9]. As the aqueous solution is not acidic, minimal to no hazardous hydrazoic acid is released, unlike other protocols. The reaction is catalyzed by zinc Lewis acids such as $ZnCl_2$, $Zn(ClO_4)_2$, and $ZnBr_2$, the latter of which was found to give the highest yields. Due to the low reactivity of acetonitrile, an electron-rich nitrile, elevated temperatures (170 °C) are required, necessitating the use of a pressure tube or autoclave as the reaction vessel. Minor optimizations increased the yield from 75% to 90% by adjusting the stoichiometry from a 1:1.1 molar ratio of nitrile to azide to a 2:1 molar ratio, as well as keeping the amount of water during workup to a minimum. Given the high water solubility of 5-methyltetrazole, minimizing water content during the workup was crucial to enable efficient extraction. This was achieved using concentrated aqueous NaOH to precipitate $Zn(OH)_2$ and concentrated aqueous HCl to protonate the sodium tetrazolate [9,12].

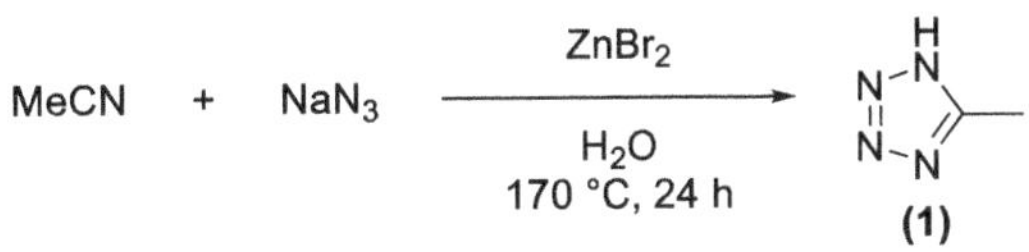

Scheme 1. Synthesis of 5-methyltetrazole.

The subsequent oxidation of 5-methyltetrazole is carried out using a fourfold excess of Oxone® in water at 40 °C (Scheme 2). Oxone® and tri-sodium phosphate dodecahydrate are added alternately over one hour, maintaining a pH of 7–8. No conversion was observed under unbuffered conditions. The maximum yield is achieved after three days. Increasing the reaction time or adding more equivalents of Oxone® at the start or after three days did not increase the yield.

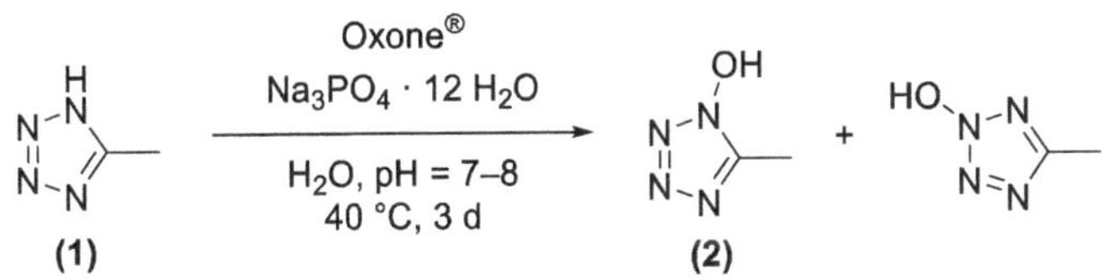

Scheme 2. Oxidation of 5-methyltetrazole.

The reaction yields the regioisomers 1-hydroxy-5-methyltetrazole (**2**) and 2-hydroxy-5-methyltetrazole (Scheme 2) in a 3:1 molar ratio, as determined by ^{1}H NMR spectroscopy. Pure compound **2** is obtained by recrystallization using a mixture of acetone and *i*-hexane in a 52% yield.

A series of high-nitrogen salts (compounds **3**–**7**) were prepared by reacting compound **2** with the appropriate bases in water, ethanol, or a mixture of both solvents (Scheme 3). For salts **4**, **5**, and **7**, equimolar amounts of base were used, whereas salt **3** was obtained using an excess of 2 M aqueous ammonia solution. The guanidinium salt **6** was synthesized by treating compound **2** with guanidinium carbonate. In the case of (D) and (E), heating the solution to 80 °C for 15 min was needed for conversion. Following solvent evaporation, all salts were obtained in good to quantitative yields. Detailed experimental procedures are found in the Supplementary Materials.

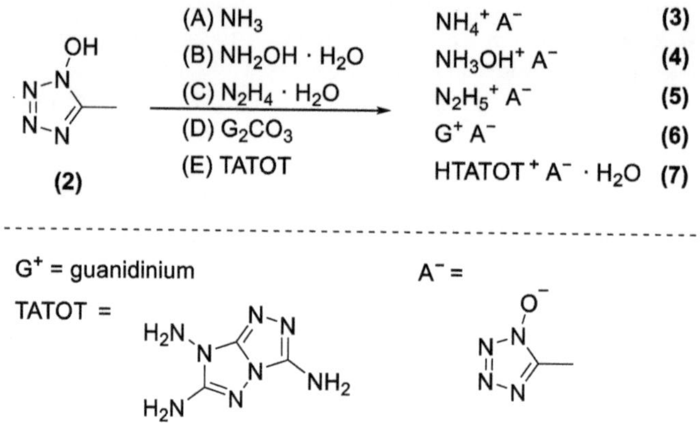

Scheme 3. Synthesis of energetic salts of 1-hydroxy-5-methyltetrazole.

2.2. X-Ray Diffraction

Single crystals suitable for X-ray diffraction were obtained by recrystallization from ethanol or water. Detailed information regarding the measurement and refinement of all compounds is provided in the Supplementary Materials (Tables S1 and S2). The crystallographic data has been deposited in the Cambridge Structural Database (CSD) under CCDC deposition numbers 2,453,219 (for **2**), 2,453,217 (for **3**), 2,453,215 (for **4**), 2,453,218 (for **5**), 2,453,216 (for **6**), and 2,453,214 (for **7**).

Compound **2** crystallizes in the orthorhombic space group Pna2_1 with a cell volume of 445.62(3) Å^3 and four molecular units per cell. The cell constants are a = 9.6928(4) Å, b = 4.0112(2) Å, and c = 11.4614(5) Å. The recalculated density at 298 K is 1.465 g cm^{-3} (Figure 1A). Compared to 5-methyltetrazole (1.349 g cm^{-3} at 296 K), the introduction of the hydroxy group results in an increase of over 0.1 g cm^{-3} [34]. The molecule is almost planar, and the N-hydroxy function bends slightly out of plane, O1–N1–C1–N4 175.76°. The bond lengths in the aromatic ring (N1–N2 1.343 Å, N2–N3 1.286 Å, and N4–C1 1.318 Å), as well as the N-bonded hydroxide (O1–N1 1.356 Å), fall within the standard range. The molecules form strong hydrogen bonds, with an H1–N4 distance of 1.63(5) Å, leading to the formation of zig-zag chains. These chains are arranged alternately in the a–c plane, forming layers that stack along the b axis (Figure 1B).

Compound **3** crystallizes in the orthorhombic space group Pbca with a cell volume of 2131.1(2) Å^3 and sixteen molecular units per cell. The cell constants are a = 13.1916(8) Å, b = 7.5481(4) Å, and c = 21.4023(10) Å, while the density is 1.460 g cm^{-3} at 298 K (Figure 2A). Compared to ammonium 5-methyltetrazolate (1.259 g cm^3 at 298 K), the density is significantly higher [35]. The bend of the N-hydroxy function of the neutral compound **2** is lost, resulting in a planar structure, with the O1–N1–N2–N3 torsion angle measuring 179.97°. In all the salts **3**–**7**, the N–O bond is shortened by approximately 2 pm compared to the neutral compound. Each ammonium cation is surrounded by four or five hydroxytetrazolate anions, forming three hydrogen bonds to neighboring oxygen atoms in the range of O1–H5C 1.75 Å to O2–H5A 1.98 Å (Figure 2B).

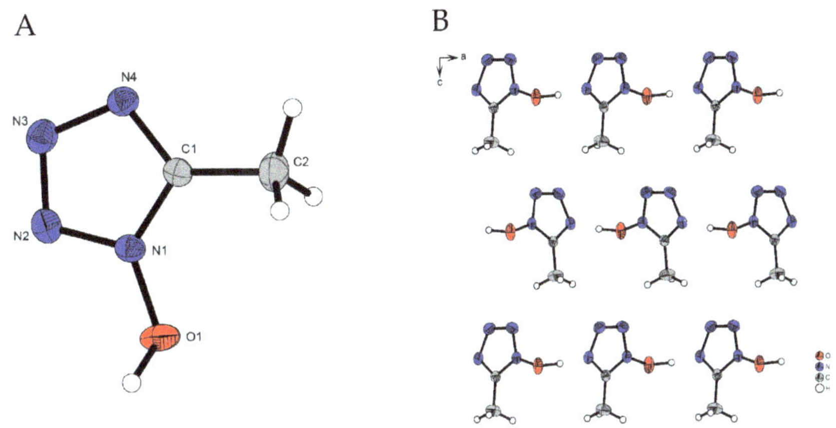

Figure 1. (**A**) Molecular structure of **2** as determined by low-temperature X-ray diffraction with thermal ellipsoids drawn at a 50% probability level. Selected bond lengths [Å], bond angles [°], and torsion angles [°]: O1–N1 1.356(2), N1–N2 1.343(3), N2–N3 1.286(3), N4–C1 1.318(2), N1–N2–N3 105.76(17), N4–C1–N1 106.15(17), N2–N1–C1–N4 0.7(2), N1–N2–N3–N4 1.2(2), and O1–N1–C1–C2 −4.6(3); (**B**) 2D layers along the *b* axis.

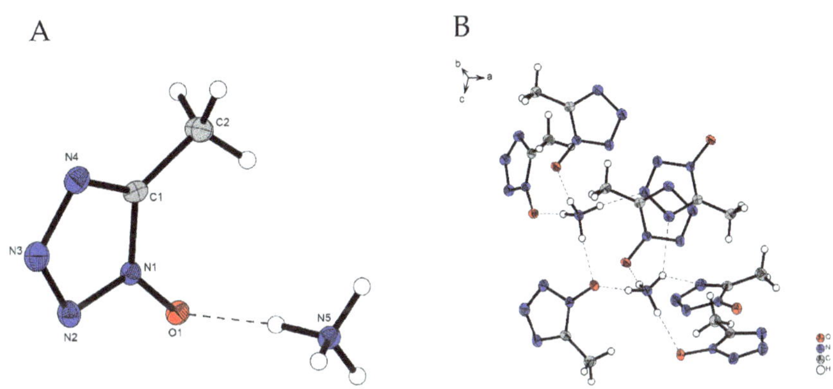

Figure 2. (**A**) Molecular structure of **3** as determined by low-temperature X-ray diffraction with thermal ellipsoids drawn at a 50% probability level. Selected bond lengths [Å], bond angles [°], and torsion angles [°]: O1–N1 1.337(18), N1–N2 1.344(2), N2–N3 1.311(2), N3–N4 1.359(2), O1–N1–C1 128.33(14), N3–N4–C1 106.14(14), O1–N1–N2–N3 179.97(13), O1–N1–C1–C2 −1.5(3), and N1–N2–N3–N4 0.0(2); (**B**) 3D structure of **3**.

Compound **4** crystallizes in the triclinic space group P-1 with a cell volume of 594.11(8) $Å^3$ and four molecular units per cell. The cell constants are a = 7.3821(5) Å, b = 8.2511(6) Å, and c = 10.2099 Å, while the density is 1.461 g cm^{-3} at 298 K (Figure 3A). The molecules form chains in the b-c plane, with hydroxylammonium cations bridging two 5-methylhydroxytetrazolate anions via hydrogen bonds (O1–H2 1.682 Å; N4–H5C 1.968 Å). The two chains are arranged at a 180° rotation, with the methyl groups facing outwards, as they do not interact as much as the hydrogen bonds (N8–H5A 2.128 Å). The layers stack closely along the b axis due to strong hydrogen bonding (O1–H5B 2.128 Å) (Figure 3B).

Compound **5** crystallizes in the orthorhombic space group Pbca with a cell volume of 2436.0(7) $Å^3$ and sixteen molecular units per cell. The cell constants are a = 14.632(2) Å, b = 6.9287(11) Å, and c = 24.028(5) Å, while the density is 1.404 g cm^{-3} at 298 K (Figure 4).

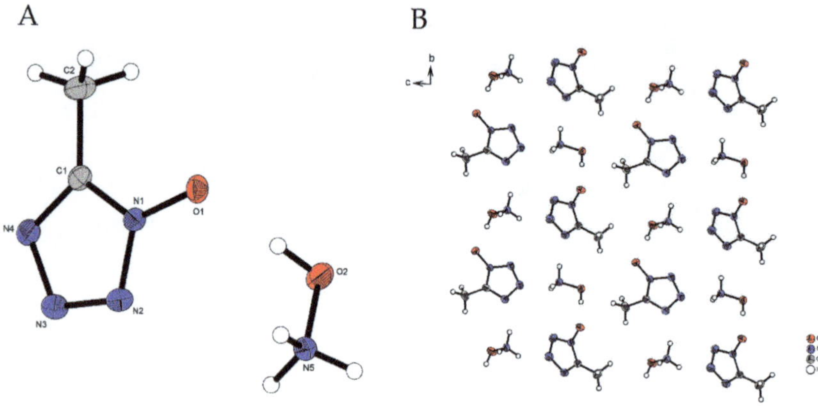

Figure 3. (**A**) Molecular structure of **4** as determined by low-temperature X-ray diffraction with thermal ellipsoids drawn at a 50% probability level. Selected bond lengths [Å], bond angles [°], and torsion angles [°]: O1–N1 1.336(15), O2–N5 1.411(17), C1–C2 1.479(2), N2–N3 1.308(18), O1–H5B 2.128 Å, O1–H2 1.682, N4–H5C 1.968, N8–H5a 2.128, N5–O2–H2 103.1(16), O1–N1–N2 122.40(11), O1–N1–N2–N3 −179.98(11), and C1–N4–N3–N2 0.37(16); (**B**) 2D layers along the *a* axis.

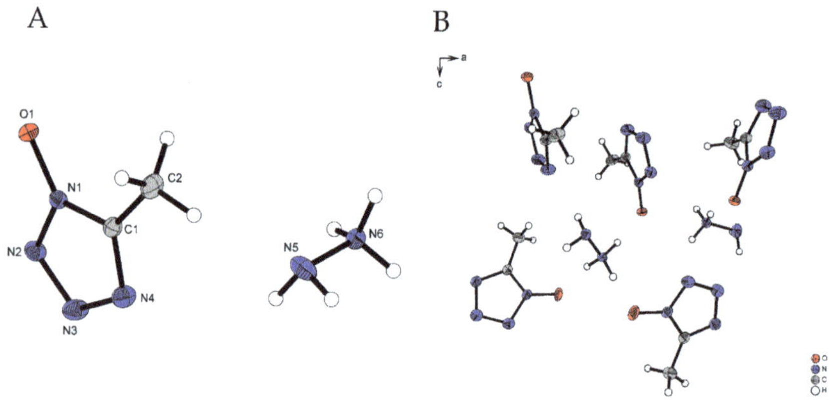

Figure 4. (**A**) Molecular structure of **5** as determined by low-temperature X-ray diffraction with thermal ellipsoids drawn at a 50% probability level. Selected bond lengths [Å], bond angles [°], and torsion angles [°]: C1–N1 1.340(2), C1–C2 1.486(3), N2–N3 1.313(2), N5–N6 1.449(2), O1–N6 1.88(3), O1–N1–N2 121.96(14), N1–C1–C2 124.60(16), O1–N1–C1–N4 179.32(16), N3–N4–C1–N1 −0.1(2), and O1–N1–C1–C2 −2.0(3); (**B**) 3D structure along the *b* axis.

Compound **6** crystallizes in the monoclinic space group C2/c with a cell volume of 1516.87(16) $Å^3$ and eight molecular units per cell. The cell constants are a = 11.9689(8) Å, b = 11.2025(7) Å, and c = 11.6752(6) Å, while the density is 1.367 g cm^{-3} at 298 K, the lowest density of all the herein investigated compounds (Figure 5A). Compared to guanidinium 5-methyltetrazolate (1.276 g cm^{-3} at 298 K), the introduction of the hydroxy group results in an increase of around 0.1 g cm^{-3} [36]. Each of the 5-methylhydroxytetrazolate anions is surrounded by four guanidinium cations, forming a network of medium to strong hydrogen bonds (O1–H5A 1.932 Å, N2–H5B 2.349 Å) (Figure 5B).

Compound **7** crystallizes with the inclusion of one water molecule in the triclinic space group P-1 with a cell volume of 569.36(11) $Å^3$ and two molecular units per cell. The cell constants are a = 6.4887(7) Å, b = 6.9104(8) Å, and c = 13.2806(14) Å, while the density is 1.547 g cm^{-3} at 298 K, the highest density of all investigated compounds (Figure 6A). In the cation, the presence of different hybridized amino groups is evident. The C-bonded amines (C–NH_2) are sp^2 hybridized, as evidenced by the trigonal planar orientation of the protons (H12A–N12–H12B 118.4°) and the shorter C–N bond lengths (N11–C4 1.327 Å and N12–C5 1.325 Å). This can be attributed to the donation of electron density by the free-electron pair into the triazole rings. The N-bonded amine (N–NH_2) is sp^3 hybridized, with the free-electron pair located at the nitrogen N10. The hydrogen atoms form an almost ideal tetrahedral angle (H10A–N10–H10B 109.6°) with a longer bond length (N7–N10 1.405 Å). The molecules form chains where HTATOT cations bridge two 5-methylhydroxytetrazolate anions through hydrogen bonds (N4–H11B 2.11 Å, O1–H5 1.69 Å, and N2–H12A 2.20 Å). Two chains are arranged in an antiparallel fashion, rotated by 180°, with the methyl- and sp^3-hybridized NH_2 groups oriented outwards. The layers are staggered with the water molecules in between connecting the layers by hydrogen bonding (Figure 6B).

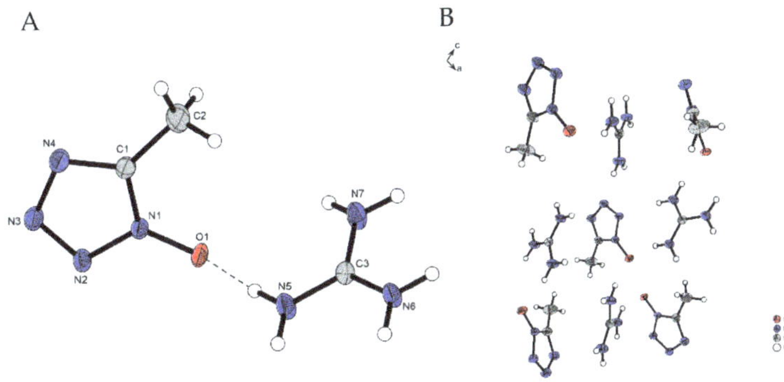

Figure 5. (**A**) Molecular structure of **6** as determined by low-temperature X-ray diffraction with thermal ellipsoids drawn at a 50% probability level. Selected bond lengths [Å], bond angles [°], and torsion angles [°]: O1–N1 1.330(15), C1–C2 1.473(3), N5–C3 1.320(2), N5–C3–N6 120.41(14), N5–C3–N7 119.60(13), N1–C1–C2 123.84(15), O1–N1–C1–C2 −2.1(2), and N3–N4–C1–N1 −0.28(15); (**B**) 3D structure along the *b* axis.

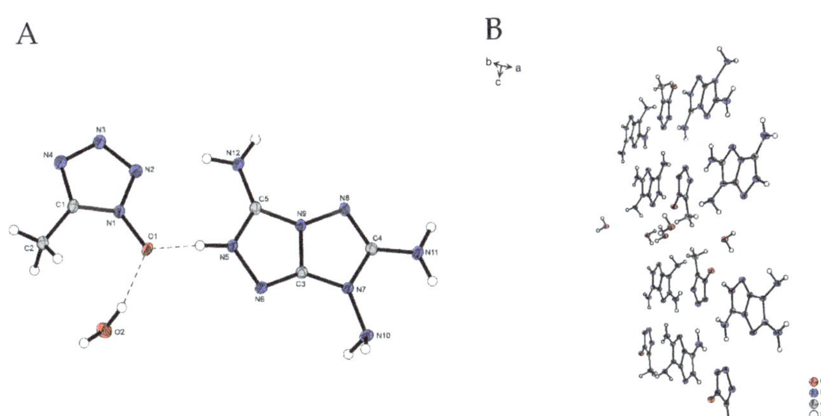

Figure 6. (**A**) Molecular structure of **7** as determined by low-temperature X-ray diffraction with thermal ellipsoids drawn at a 50% probability level. Selected bond lengths [Å], bond angles [°], and torsion angles [°]: N5–N6 1.416(19), O1–N1 1.335(17), N7–N10 1.405(18), N7–C4 1.396(2), N11–C4 1.327, C5–N12 1.325(2), N6–C3–N7 140.07(15), N5–C5–N9 104.43(13), N11–C4–N7 120.54(14), N10–N7–C3–N6 16.2(3), N8–N9–C5–N12 0.9(3), and N10–N7–C3–N9 −165.91(14); (**B**) 2D layers.

2.3. Physicochemical Properties

Impact sensitivities (ISs) and friction sensitivities (FSs) were determined according to BAM (Bundesanstalt für Materialforschung und-prüfung) standards, using a BAM drop hammer and friction apparatus, applying the 1 of 6 method [37,38]. Table 1 lists the experimental values. Except for the neutral compound **2** with 3 J and 160 N, all other investigated compounds, **3–6**, are insensitive towards impact (>40 J) and show little sensitivity towards friction with hydroxylammonium salt **4** being the most sensitive (288 N), while the others are insensitive with 360 N or >360 N.

Table 1. Physicochemical properties of the investigated compounds **2**, **3**, **4**, **5**, **6**, and **TNT**.

	2	**3**	**4**	**5**	**6**	**TNT**
Formula	$C_2H_4N_4O$	$C_2H_7N_5O$	$C_2H_7N_5O_2$	$C_2H_8N_6O$	$C_3H_9N_7O$	$C_7H_5N_3O_6$
M [g mol^{-1}]	100.08	117.11	133.11	132.13	159.15	227.13
IS [J] (a)	3	>40	>40	>40	>40	15
FS [N] (b)	160	360	288	360	>360	>360
ρ (298 K) [g cm^{-3}] (c)	1.465	1.460	1.461	1.404	1.367	1.65
N [%] (d)	55.9	59.8	52.6	63.6	61.6	18.5
Ω [%] (e)	−48.0	−61.5	−42.1	−60.6	−65.4	−24.7
T_{endo} [°C] (f)	146	166	139	105	68, 182	80
T_{dec} [°C] (g)	194	229	141	224	256	290
$\Delta_f H°$ [kJ mol^{-1}] (h)	230.5	177.5	245.0	335.6	164.4	−59
Explo5 V7.01.01						
$-\Delta_{Ex} U^0$ [kJ kg^{-1}] (i)	4483	3966	5393	4823	3008	4363
T_{det} [K] (j)	2966	2459	3217	2817	2098	3165
V_0 [L kg^{-1}] (k)	776	859	867	884	593	593
P_{CJ} [kbar] (l)	186	212	230	219	156	186
V_{det} [m s^{-1}] (m)	7343	7982	8065	8109	7090	6817

(a) Impact sensitivity (BAM drop hammer, 1 of 6) [37,38]; (b) friction sensitivity (BAM friction tester, 1 of 6) [37,38]; (c) density from X-ray diffraction analysis recalculated to 298 K; (d) nitrogen content; (e) oxygen balance with respect to CO; (f) temperature of endothermic event (DTA; β = 5 °C min^{-1}); (g) decomposition temperature (DTA; β = 5 °C min^{-1}); (h) calculated (CBS-4 M) heat of formation; (i) energy of explosion; (j) explosion temperature; (k) volume of detonation products (assuming only gaseous products); (l) detonation pressure at the Chapman–Jouguet point; (m) detonation velocity.

The thermal stabilities were determined using differential thermal analysis (DTA) at a heating rate of 5 °C min^{-1}. Thermal stabilities were determined by differential thermal analysis (DTA) with a heating rate of 5 °C min^{-1}. The guanidinium salt **6** has the highest decomposition temperature of 256 °C. 1-Hydroxy-5-methyltetrazole melts at 146 °C and begins to decompose at 194 °C, which is 60 °C lower than 5-methyltetrazole **1**, which decomposes at 254 °C [39]. Compound **4** has a melting point of 139 °C, after which it decomposes. The ammonium **3** and the hydrazinium salt **5** decompose at similar temperatures of 229 °C and 224 °C, respectively.

The EXPLO5 V7.01.01 code was used to calculate the energetic performance parameters, using X-ray densities recalculated to room temperature and solid-state heats of formation obtained via CBS-4M calculations using the Gaussian 16 program, Revisions A.04. The energetic performance improved, and the oxygen balance increased from −76% to −48% by oxidizing 5-methyltetrazole (**1**). The detonation velocity increased from 6683 m s^{-1} to 7343 m s^{-1} [34]. Further increases in the energetic parameters were achieved through salification, with compounds **3**, **4**, and **5** improving substantially to around 8000 m s^{-1}, with a detonation pressure of between 212 and 230 kbar. Only compound **6** deteriorated, with a V_{det} of 7090 m s^{-1}, though still outperforming TNT. 1-hydroxy-5-methyltetrazole has the highest density of 1.465 g cm^{-3}, while the salts range from 1.367 to 1.461 g cm^{-3}.

2.4. Toxicity

A comparative in silico toxicological assessment was conducted using the ProTox 3.0 platform to evaluate the acute toxicity (LD_{50}), mutagenicity, carcinogenicity, and acute ecotoxicity (on fish, daphnids and algae) of compounds **2–6**, TNT (2,4,6-trinitrotoluene), RDX (1,3,5-trinitro-1,3,5-triazinane), and HMX (1,3,5,7-tetranitro-1,3,5,7-tetrazocane) [40–42]. The prediction accuracy for compounds **2–6** was comparatively low (23%) relative to TNT (69%) and RDX/HMX (68%), indicating the need for further experimental validation. Predicted LD_{50} values suggest that compounds **2–5** (890 mg kg^{-1}) and especially compound **6** (3150 mg kg^{-1}) exhibit significantly lower acute toxicity compared to TNT (607 mg kg^{-1}), RDX (100 mg kg^{-1}), and HMX (186 mg kg^{-1}). All new compounds were predicted to exhibit non-acute ecotoxicity with 55–71% probability, in contrast to the commercial explosives, which were predicted to show acute ecotoxicity with similar probabilities (TNT 77%, RDX 50%, HMX 53%). Regarding genotoxicity, compounds **2–5** were predicted to be non-mutagenic (52–62% probability), while guanidinium salt **6** was predicted to be mutagenic with a 61% probability. By contrast, TNT, RDX, and HMX were all predicted to be mutagenic with a high probability (99%, 89%, and 87%, respectively). Predictions of carcinogenicity indicate that compounds **2–6** are active (with a probability of 58–69%), as are RDX and HMX (with probabilities of 86% and 87%, respectively). In contrast, TNT was predicted to be inactive (55%).

Overall, the newly synthesized compounds **2–6** demonstrate favorable properties in terms of predicted acute toxicity, ecotoxicity, and mutagenicity when compared to conventional explosives TNT, RDX, and HMX. However, they are worse in terms of predicted carcinogenicity compared to TNT. The full reports are given in the Supplementary Materials.

3. Conclusions

1-Hydroxy-5-methyltetrazole was synthesized in two steps using common chemicals and water as the reaction solvent. The [2 + 3] cycloaddition of acetonitrile with sodium azide catalyzed by zinc bromide was optimized. The subsequent oxidation of 5-methyltetrazole with Oxone® and sodium phosphate dodecahydrate as a buffer produced compound **2**. Five nitrogen-rich salts, ammonium (**3**), hydroxylammonium (**4**), hydrazinium (**5**), guanidinium (**6**), and HTATOT (**7**) were synthesized by reacting the respective bases with compound **2** in water or ethanol. The energetic properties of the 5-methyltetrazole scaffold could be drastically improved by oxidation and salification. Although the hydroxylammonium salt **4** performs well, it is unsuitable due to its low decomposition temperature of 141 °C. Guanidinium salt **6** is the most thermally stable with a decomposition temperature of 256 °C; however, its low detonation velocity of 7090 m s^{-1} renders it unsuitable as a replacement candidate. A promising candidate for a cost-effective, insensitive explosive (40 J, >360 N) is ammonium salt **3**, which has a decomposition temperature of 229 °C and a detonation velocity of 7982 m s^{-1}. The best-performing compound is hydrazinium salt **5**, with a detonation velocity of 8109 m s^{-1} and a detonation pressure of 219 kbar. It is insensitive towards friction and impact (40 J, >360 N) and has a high decomposition temperature of 224 °C.

Supplementary Materials: The following supporting information can be downloaded at https://www.mdpi.com/article/10.3390/molecules30132766/s1, Detailled information on the synthesis and characterization of the presented compounds; Tables S1 and S2: Crystallographic data and structure refinement details of the prepared compounds; Table S3: CBS-4M electronic enthalpies for atoms C, H, N, and O and their literature values; Table S4: Calculation results; ^{1}H and ^{13}C NMR spectra of all prepared compounds; reports of predicted toxicity of compounds **2–6**, TNT, RDX and HMX. References [9,17,37,38,43–58] are cited in the supplementary materials.

Author Contributions: Methodology, L.J.E. and M.B.; investigation, L.J.E. and M.B.; practical work, L.J.E. and M.B.; writing—original draft, L.J.E.; supervision, J.S. and T.M.K. All authors have read and agreed to the published version of the manuscript.

Funding: The financial support of this work by the Ludwig-Maximilian University (LMU), EMTO GmbH, BIAZZI SA, and Eurenco in France is greatly acknowledged.

Institutional Review Board Statement: Not applicable.

Informed Consent Statement: Not applicable.

Data Availability Statement: The data presented in this study can be found in the article and in the Supplementary Materials.

Acknowledgments: For their financial support of this work, Ludwig-Maximilian University (LMU), BIAZZI, and EMTO GmbH are gratefully acknowledged.

Conflicts of Interest: The authors declare no conflicts of interest.

References

1. Klapötke, T.M. *Chemistry of High-Energy Materials*, 6th ed.; De Gruyter: Berlin, Germany; Boston, MA, UAA, 2022.
2. Benz, M.; Delage, A.; Klapötke, T.M.; Kofen, M.; Stierstorfer, J. A rocky road toward a suitable TNT replacement—A closer look at three promising azoles. *Propellants Explos. Pyrotech.* **2023**, *48*, e202300042. [CrossRef]
3. Conder, J.M.; La Point, T.W.; Steevens, J.A.; Lotufo, G.R. Recommendations for the assessment of TNT toxicity in sediment. *Environ. Toxicol. Chem.* **2004**, *23*, 141–149. [CrossRef] [PubMed]
4. Bradley, P.M.; Chapelle, F.H.; Landmeyer, J.E.; Schumacher, J.G. Microbial Transformation of Nitroaromatics in Surface Soils and Aquifer Materials. *Appl. Environ. Microbiol.* **1994**, *60*, 2170–2175. [CrossRef]
5. Pichtel, J. Distribution and Fate of Military Explosives and Propellants in Soil: A Review. *Appl. Environ. Soil Sci.* **2012**, *1*, 617236. [CrossRef]
6. Bhadra, R.; Wayment, D.G.; Williams, R.K.; Barman, S.N.; Stone, M.B.; Hughes, J.B.; Shanks, J.V. Studies on plant-mediated fate of the explosives RDX and HMX. *Chemosphere* **2001**, *44*, 1259–1264. [CrossRef]
7. Anastas, P.; Eghbali, N. Green Chemistry: Principles and Practice. *Chem. Soc. Rev.* **2010**, *39*, 301–312. [CrossRef]
8. Zhang, J.; Yin, P.; Pan, G.; Wang, Z.; Zhang, J.; Mitchell, L.A.; Parrish, D.A.; Shreeve, J.N.M. 5-(4-Azidofurazan-3-yl)-1-hydroxytetrazole and its derivatives: From green primary to secondary explosives. *New J. Chem.* **2019**, *43*, 12684–12689. [CrossRef]
9. Demko, Z.P.; Sharpless, K.B. Preparation of 5-Substituted 1H-Tetrazoles from Nitriles in Water. *J. Org. Chem.* **2001**, *66*, 7945–7950. [CrossRef]
10. Vorona, S.; Artamonova, T.; Zevatskii, Y.; Myznikov, L. An Improved Protocol for the Preparation of 5-Substituted Tetrazoles from Organic Thiocyanates and Nitriles. *Synthesis* **2014**, *46*, 781–786.
11. Babu, A.; Sinha, A. Catalytic Tetrazole Synthesis via [3+2] Cycloaddition of NaN_3 to Organonitriles Promoted by Co(II)-complex: Isolation and Characterization of a Co(II)-diazido Intermediate. *ACS Omega* **2024**, *9*, 21626–21636. [CrossRef]
12. Finnegan, W.G.; Henry, R.A.; Lofquist, R. An Improved Synthesis of 5-Substituted Tetrazoles. *J. Am. Chem. Soc.* **1958**, *80*, 3908–3911. [CrossRef]
13. Wang, T.; Xu, L.; Dong, J. FSO_2N_3-Enabled Synthesis of Tetrazoles from Amidines and Guanidines. *Org. Lett.* **2023**, *25*, 6222–6227. [CrossRef]
14. Stepanov, A.I.; Sannikov, V.S.; Dashko, D.V.; Roslyakov, A.G.; Astrat'ev, A.A.; Stepanova, E.V. A new preparative method and some chemical properties of 4-R-furazan-3-carboxylic acid amidrazones. *Chem. Heterocycl. Compd.* **2015**, *51*, 350–360. [CrossRef]
15. Oberhummer, W. Die Reaktion aliphatischer Iminoäther mit Hydrazin. *Monatsh. Chem.* **1933**, *63*, 285–300. [CrossRef]
16. Zhang, J.-G.; Li, J.-Y.; Zang, Y.; Shu, Y.-J.; Zhang, T.-L.; Yang, L.; Power, P.P. Synthesis and Characterization of a Novel Energetic Complex $[Cd(DAT)_6](NO_3)_2$ (DAT = 1,5-diamino-tetrazole) with High Nitrogen Content. *Z. Anorg. Allg. Chem.* **2010**, *636*, 1147–1151. [CrossRef]
17. Bauer, L.; Benz, M.; Klapötke, T.M.; Pignot, C.; Stierstorfer, J. Combining the most suitable energetic tetrazole and triazole moieties: Synthesis and characterization of 5-(1-hydroxy-3-nitro-1,2,4-triazol-5-yl)-1-hydroxy-tetrazole and its nitrogen-rich ionic derivatives. *Mater. Adv.* **2022**, *3*, 3945–3951. [CrossRef]
18. Klapötke, T.M.; Kofen, M.; Schmidt, L.; Stierstorfer, J.; Wurzenberger, M.H.H. Selective Synthesis and Characterization of the Highly Energetic Materials 1-Hydroxy-5H-tetrazole (CHN4O), its Anion 1-Oxido-5H-tetrazolate (CN_4O^-) and Bis(1-hydroxytetrazol-5-yl)triazene. *Chem. Asian J.* **2021**, *16*, 3001–3012. [CrossRef]

19. Klapötke, T.M.; Piercey, D.G.; Stierstorfer, J. The Taming of CN_7^-: The Azidotetrazolate 2-Oxide Anion. *Chem. Eur. J.* **2011**, *17*, 13068–13077. [CrossRef]
20. Fischer, N.; Fischer, D.; Klapötke, T.M.; Piercey, D.G.; Stierstorfer, J. Pushing the limits of energetic materials—The synthesis and characterization of dihydroxylammonium 5,5′-bistetrazole-1,1′-diolate. *J. Mater. Chem.* **2012**, *22*, 20418–20422. [CrossRef]
21. Begtrup, M.; Vedsø, P. Preparation of N-hydroxyazoles by oxidation of azoles. *J. Chem. Soc. Perkin Trans.* **1995**, *1*, 243–247. [CrossRef]
22. Giles, R.G.; Lewis, N.J.; Oxley, P.W.; Quick, J.K. Facile synthesis of novel 2-hydroxytetrazole derivatives via selective N-2 oxidation. *Tetrahedron Lett.* **1999**, *40*, 6093–6094. [CrossRef]
23. Göbel, M.; Karaghiosoff, K.; Klapötke, T.M.; Piercey, D.G.; Stierstorfer, J. Nitrotetrazolate-2N-oxides and the Strategy of N-Oxide Introduction. *J. Am. Chem. Soc.* **2010**, *132*, 17216–17226. [CrossRef]
24. Plenkiewicz, J. Rearrangement of azidoximes to tetrazole derivatives. *Tetrahedron Lett.* **1975**, *16*, 341–342. [CrossRef]
25. Lal, S.; Staples, R.J.; Shreeve, J.N.M. Design and synthesis of phenylene-bridged isoxazole and tetrazole-1-ol based energetic materials of low sensitivity. *Dalton Trans.* **2023**, *52*, 3449–3457. [CrossRef]
26. Zhao, S.; Ali, A.S.; Kong, X.; Zhang, Y.; Liu, X.; Skidmore, M.A.; Forsyth, C.M.; Savage, G.P.; Wu, D.; Xu, Y.; et al. 1-Benzyloxy-5-phenyltetrazole derivatives highly active against androgen receptor-dependent prostate cancer cells. *Eur. J. Med. Chem.* **2023**, *246*, 114982. [CrossRef]
27. Zhang, J.; Jin, Z.; Hao, W.; Guo, Z.; Wang, Y.; Peng, R.; Jin, B. $N_2H_6^{2+}$ Energetic Salts: A Promising Strategy for Increasing the Density and Balance Safety and Energy of Explosives. *Cryst. Growth Des.* **2023**, *23*, 4499–4505. [CrossRef]
28. Bettinetti, G.F.; Maffei. The action of hydrazoic acid on nitrolic acids. *S. Ann. Chim.* **1956**, *46*, 812–815.
29. Eberhardt, L.J.; Benz, M.; Stierstorfer, J.; Klapötke, T.M. Synthesis and Characterization of 1-Hydroxy-5-Methyltetrazole and its Energetic Salts. In Proceedings of the 27th Seminar New Trends in Research of Energetic Materials, Pardubice, Czech Republic, 2–4 April 2025.
30. Mihina, J.S.; Herbst, R.M. The Reaction of Nitriles with Hydrazoic Acid: Synthesis of Monosubstituted Tetrazoles. *J. Org. Chem.* **1950**, *15*, 1082–1092. [CrossRef]
31. Abrishami, F.; Ebrahimikia, M.; Rafiee, F. Synthesis of 5-substituted 1H-tetrazoles using a recyclable heterogeneous nanonickel ferrite catalyst. *Appl. Organomet. Chem.* **2015**, *29*, 730–735. [CrossRef]
32. Sudhakar, K.; Rao, B.P.C.; Kumar, B.P.; Suresh, M.; Ravi, S. Syntheses of 5-Substituted 1H-tetrazoles Catalyzed by Reusable Cu(II)-NaY Zeolite from Nitriles. *Asian J. Chem.* **2017**, *29*, 864–866. [CrossRef]
33. Movaheditabar, P.; Javaherian, M.; Nobakht, V. CuO/aluminosilicate as an efficient heterogeneous nanocatalyst for the synthesis and sequential one-pot functionalization of 5-substituted-1H-tetrazoles. *React. Kinet. Mech. Cat.* **2017**, *122*, 217–228. [CrossRef]
34. Ohno, Y.; Akutsu, Y.; Arai, M.; Tamura, M.; Matsunaga, T. 5-Methyl-1H-tetrazole. *Acta Crystallog. C* **1999**, *55*, 1014–1016. [CrossRef]
35. Zhang, X.-M.; Zhao, Y.-F.; Wu, H.-S.; Batten, S.R.; Ng, S.W. Syntheses and structures of metal tetrazole coordination polymers. *Dalton Trans.* **2006**, *26*, 3170–3178. [CrossRef] [PubMed]
36. Inoue, K.; Sasahara, Y.; Matsumoto, S.; Kumasaki, M. Synthesis, crystallographic characterization, and thermal analyses of five 5-methyl-1H-tetrazole salts. *Sci. Technol. Energ. Ma.* **2024**, *85*, 53–59.
37. Reichel & Partner GmbH. Available online: https://www.rp-mespro.de/de (accessed on 20 June 2025).
38. United Nations. *Test Methods According to the UN Recommendations on the Transport of Dangerous Goods, Manual of Test and Criteria*, 4th ed.; United Nations Publication: New York, NY, USA; Geneva, Switzerland, 2003.
39. Shen, C.; Xu, Y.-G.; Lu, M. A series of high-energy coordination polymers with 3,6-bis(4-nitroamino-1,2,5-oxadiazol-3-yl)-1,4,2,5-dioxadiazine, a ligand with multi-coordination sites, high oxygen content and detonation performance: Syntheses, structures, and performance. *J. Mater. Chem. A* **2017**, *5*, 18854–18861. [CrossRef]
40. Banerjee, P.; Kemmler, E.; Dunkel, M.; Preissner, R. ProTox 3.0: A webserver for the prediction of toxicity of chemicals. *Nucleic Acids Res.* **2024**, *52*, 513–520. [CrossRef]
41. Banerjee, P.; Eckert, A.O.; Schrey, A.K.; Preissner, R. ProTox-II: A webserver for the prediction of toxicity of chemicals. *Nucleic Acids Res.* **2018**, *46*, 257–263. [CrossRef]
42. ProTox 3.0. Available online: https://tox.charite.de/protox3/ (accessed on 26 June 2025).
43. *CrysAlisPro Software*; Version 171.33.41; Rigaku Corporation: Wroclaw, Poland, 2009. Available online: http://www.rigaku.com (accessed on 20 June 2025).
44. Sheldrick, G.M. SHELXT Integrated space-group and structure determination. *Acta Cryst.* **2015**, *A71*, 3–8. [CrossRef]
45. Dolomanov, O.V.; Bourhis, L.J.; Gildea, R.J.; Howard, J.A.K.; Puschmann, H. OLEX2: A complete structure solution, refinement and analysis program. *J. Appl. Crystallogr.* **2009**, *42*, 339–341. [CrossRef]
46. *SCALE3 ABSPACK—An Oxford Diffraction Program*; Version 1.0.4; Rigaku Corporation: Wroclaw, Poland, 2009.
47. *APEX3*; Version 2018.7-2; Bruker AXS Inc.: Madison, WI, USA, 2018.

48. Frisch, M.J.; Trucks, G.W.; Schlegel, H.B.; Scuseria, G.E.; Robb, M.A.; Cheeseman, J.R.; Scalmani, G.; Barone, V.; Petersson, G.A.; Nakatsuji, H.; et al. *Gaussian 16*; Revision A.04; Gaussian, Inc.: Wallingford, CT, USA, 2016.
49. Ochterski, J.W.; Petersson, G.A.; Montgomery, J.A., Jr. A complete basis set model chemistry. V. Extensions to six or more heavy atoms. *J. Chem. Phys.* **1996**, *104*, 2598–2619. [CrossRef]
50. Montgomery, J.A.; Frisch, M.J.; Ochterski, J.W.; Petersson, G.A. A complete basis set model chemistry. VII. Use of the minimum population localization method. *J. Chem. Phys.* **2000**, *12*, 6532–6542.
51. Curtiss, L.A.; Raghavachari, K.; Redfern, P.C.; Pople, J.A. Assessment of Gaussian-2 and density functional theories for the computation of enthalpies of formation. *J. Chem. Phys.* **1997**, *106*, 1063–1079. [CrossRef]
52. Byrd, E.F.C.; Rice, B.M. Improved Prediction of Heats of Formation of Energetic Materials Using Quantum Mechanical Calculations. *Phys. Chem.* **2006**, *110*, 1005–1013. [CrossRef]
53. Rice, B.M.; Pai, S.V.; Hare, J. Predicting heats of formation of energetic materials using quantum mechanical calculations. *Comb. Flame* **1999**, *118*, 445–458. [CrossRef]
54. Lindstrom, P.J.; Mallard, W.G. NIST Standard Reference Database Number 69. Available online: http://webbook.nist.gov/chemistry/ (accessed on 26 June 2020).
55. Westwell, M.S.; Searle, M.S.; Wales, D.J.; Williams, D.H.; Trouton, F. Empirical Correlations between Thermodynamic Properties and Intermolecular Forces. *J. Am. Chem. Soc.* **1995**, *117*, 5013–5015. [CrossRef]
56. Trouton, F. IV. On molecular latent heat. *Philos. Mag.* **1884**, *18*, 54–57. [CrossRef]
57. Jenkins, H.D.B.; Roobottom, H.K.; Passmore, J.; Glasser, L. Relationships among Ionic Lattice Energies, Molecular (Formula Unit) Volumes, and Thermochemical Radii. *Inorg. Chem.* **1999**, *38*, 3609–3620. [CrossRef]
58. Jenkins, H.D.B.; Tudela, D.; Glasser, L. Lattice potential energy estimation for complex ionic salts from density measurements. *Inorg. Chem.* **2002**, *41*, 2364–2367. [CrossRef]

molecules

MDPI

Article

Thermal Interaction Mechanisms of Ammonium Perchlorate and Ammonia Borane

Yunlong Zhang, Rui Pu, Shaoli Chen and Qilong Yan *

National Key Laboratory on Solid Rocket Propulsion, Northwestern Polytechnical University, Xi'an 710072, China
* Correspondence: qilongyan@nwpu.edu.cn; Tel.: +029-88494031

Abstract: Ammonia borane (AB), with a theoretical hydrogen content of 19.6 wt%, is constrained by its low crystalline density (0.758 g/cm^3) and poor thermal stability (decomposing at 100 °C). In this study, AB/ammonium perchlorate (AP) composites were synthesized via freeze-drying at a 1:1 molar ratio. The integration of AP introduced intermolecular interactions that suppressed AB decomposition, increasing the onset temperature by 80 °C. Subsequent vacuum calcination at 100 °C for 2 h formed oxygen/fuel-integrated ammonium perchlorate borane (APB), which achieved decomposition temperatures exceeding 350 °C. The proposed mechanism involved AB decomposing into borazine and BN polymers at 100 °C, which then $NH_3BH_2^+/ClO_4^-$ combined to form APB. At 350 °C, APB underwent the following redox reactions: $4NH_3BH_2ClO_4 \rightarrow N_2\uparrow + 4HCl\uparrow + 2B_2O_3 + N_2O\uparrow + O_2\uparrow + 7H_2O\uparrow + H_2\uparrow$, while residual AP decomposed. The composite exhibited improved density (1.66 g/cm^3) and generated H_2, N_2, O_2, and HCl, demonstrating potential for hydrogen storage. Additionally, safety was enhanced by the suppression of AB's exothermic decomposition (100–200 °C). APB, with its high energy density and thermal stability, was identified as a promising high-energy additive for high-burning-rate propellants.

Keywords: ammonia borane; ammonium perchlorate; $NH_3BH_2ClO_4$; integrated design of fuel oxidant

1. Introduction

Solid propellants served as the power source for various solid motors used in missiles, weaponry, and spacecraft for space launches. The energy performance of propellants determines their effective range and deterrence capability. Consequently, recent research in the field of high-energy solid propellants has focused primarily on exploring and developing new high-energy-density materials and enhancing the energy content of existing energetic materials. Hydrogen storage fuels [1] release hydrogen upon thermal decomposition, which reduces the average molecular weight of combustion products, thereby increasing the specific impulse (I_{sp}) of the propellant. Therefore, incorporating hydrogen storage materials into propellant compositions could effectively enhance the energy level of solid propellants [1]. Chemical hydride hydrogen storage materials mainly refer to hydrogen combined with other solid materials, such as metals, through ionic or coordination bonds. Chemical hydrogen storage is the most extensively researched and promising method of hydrogen storage. It is broadly categorized into metal-based and non-metal-based hydrogen storage materials. Metal-based hydrogen storage fuels primarily include metal hydrides such as aluminum hydride (AlH_3) and metal borohydrides [2–14]. However, AlH_3 is unstable and reacts with water and moist air to produce hydrogen, while borohydrides have high thermodynamic stability, which poses challenges in hydrogen release during use.

https://doi.org/10.3390/molecules30132680

Ammonia borane (AB), as a non-metal hydrogen storage fuel [7,15], has garnered significant attention due to its high chemical hydrogen storage capacity (theoretical hydrogen content of 19.6 wt%). The first weight loss of AB occurs during the melting process, which is caused by the thermal dehydrogenation of AB, accompanied by the generation of foamy products [16]. This exothermic process is very rapid and reaches the maximum decomposition rate between 107 and 125 °C, depending on factors such as heating rate, impurities, and particle size [17–23]. The second decomposition step occurs at approximately 130 °C, accompanied by hydrogen release [24]. In the temperature range of 95 °C to 177 °C, due to the slow decomposition reaction, two exothermic peaks are typically observed on the DSC curve. After the two-step decomposition, a polymeric residue $(BNH)_x$ is obtained, which further dehydrogenates at temperatures well above 500 °C [17–23]. When the temperature is raised to 1170–1500 °C, the product transforms into hexagonal boron nitride crystals.

Other by-products have been detected through TGA-MS or TGA-FTIR techniques, primarily borazine ($B_3N_3H_6$) [25,26]. During the first decomposition step, approximately 7 wt% of $B_3N_3H_6$ is formed within the temperature range of 100–200 °C, with the remainder being produced in the second step [27]. Other minor gaseous products included diborane, ammonia, H_2N-BH_2, $H_2BNH_2BH_3$, and sublimated AB.

From these observations, three distinct reaction pathways for the thermal decomposition of AB have been summarized. The hydrogen release process of AB encompasses the three stages mentioned above, accompanied by the release of hydrogen gas and impurities, as well as the generation of $(BNH)_x$, which is mainly produced during the second stage. Due to the polarity differences between the B-H and N-H bonds in AB, the thermal decomposition of AB proceeds through anisotropic interactions and intermolecular dehydrogenation bonding [28–30].

In the initial stages of decomposition, chain products such as BH_2, BH, and BH_3 are rapidly formed. These compounds subsequently transform into diammoniate of diborane (DADB) and H_2 [30–36]. During this process, by-products such as ammonia diborane (ADB) are also produced, accompanied by the release of a significant amount of hydrogen. The chain substances undergo autocatalytic reactions to form polyaminoboranes (PABs). Under the catalysis of AB and NH_2BH_2, PABs participate in two competing reactions: propagation to form acyclic PABs or cyclization. Due to subsequent cyclization and cross-linking reactions, the formation of cyclic PABs requires higher temperatures, and these reactions also lead to the formation of borazine (BZ) and polyborazylene (PBZ). The presence of NH_3 can facilitate the reverse synthesis of AB from $NH_3BH_2(\mu\text{-}H)BH_3$ [37,38]. DADB is the initial product of further decomposition of cyclic products of AB and can also be formed through dehydrogenation bonding between AB molecules. The ring-opening reaction of DADB leads to its gradual conversion into BZ, ultimately yielding polyborazylene (PBZ).

Notably, some progress has been made in the research on AB with oxidizers. Studies found that when AB is composited with AP, the presence of NH_4^+ ions allows the thermal reaction of AB to follow an alternative pathway, generating $[NH_3BH_3NH_3]^+[ClO_4]^-$ instead of DADB [28–36,39,40].

AB is an attractive hydrogen storage material, not only due to its high theoretical hydrogen storage capacity but also because of its easier hydrogen release, which outperforms other hydrogen storage materials such as AlH_3 and metal borohydrides [8–12]. However, its low density and poor thermal stability limit its applications. When AB is composited with AP, the presence of NH_4^+ ions enables the thermal reaction of AB to follow an alternative pathway, generating $[NH_3BH_2NH_3]^+[ClO_4]^-$ instead of DADB [28–36,40], which indicates the potential formation of ammonium perchlorate borane (APB).

Guillaume [41] successfully synthesized high-energy-density ammonia dinitroamine borane through the reaction of dinitroamine with ammonia borane. This synthesis method provides a research idea to solve the problem of low density and low thermal stability of AB.

In view of the poor thermal stability and low density of AB, the researchers used AB, AP, graphene (GA), and graphene oxide (GO) as raw materials, weighed AB and AP according to the molar ratio of 1:1, and prepared AB/AP composite particles by the direct freeze-drying method. At the same time, 2 wt% GA or GO was added to prepare carbon material modified AB/AP-GA and AB/AP-GO. Then, AB/AP(Δ), AB/AP-GA(Δ), and AB/AP-GO(Δ) were synthesized by thermal reaction of the three under vacuum. The specific formula is shown in Tables 1 and 2. APB was detected in three kinds of calcined AB/AP. The pyrolysis mechanism and physicochemical properties of APB were systematically studied [28–36,40]. The prepared oxidant/fuel integrated APB has high thermal stability and high density. APB has excellent energy performance and thermal stability, and is considered to be a promising high-energy additive for high-burning-rate solid propellant.

Table 1. Composition of composite particles.

Weight (%)	AB	AP	GA	GO
AB/AP	50	50	-	-
AB/AP-GA	49	49	2	-
AB/AP-GO	49	49	-	2

Table 2. Preparation process of calcinated AB/AP.

Calcinated AB/AP	Preparation Process
AB/AP(Δ)	Calcinate AB/AP at 100 °C for two h
AB/AP-GA(Δ)	Calcinate AB/AP-GA at 100 °C for two h
AB/AP-GO(Δ)	Calcinate AB/AP-GO at 100 °C for two h

In response to AB's poor thermal stability and low density, researchers prepared AB/AP composite particles via direct freeze-drying and thermally reacted them under vacuum to synthesize APB. The pyrolysis mechanism and physicochemical properties of APB were systematically studied. The resulting oxidant/fuel-integrated APB exhibited both high thermal stability and increased density. Due to its superior energy performance and excellent thermal stability, APB was considered a promising high-energy additive for high-burning-rate solid propellants.

2. Results

2.1. Microscopic Morphology and Elemental Compositions

SEM morphological analysis revealed that AB particles exhibited irregular shapes, with most particles measuring approximately 100 μm and displaying numerous surface pits. In contrast, AP crystals predominantly adopted regular cubic or spherical morphologies, with an average particle size similarly around 100 μm. Figure 1 shows the morphology of the AB/AP mechanical mixture, where well-defined AP particles and AB particles with surface cavities and cracks were observed, both with particle sizes close to 100 μm. The AB/AP crystal morphology, as depicted in Figure 1, presented a fragmented structure alongside agglomerates of regular crystals. Notably, small, well-defined particles appeared enveloped by fibrous, cotton-like particles, with an average crystal size of only 30 μm. AB/AP-GA displayed agglomerative growth characteristics, featuring elongated graphene strands embedded within the composite particles. The overall particle morphology resembled that of AB/AP, with agglomerated particles measuring between 30 and

40 μm. Similarly, AB/AP-GO demonstrated crystal agglomeration but with larger dimensions, accompanied by numerous small, flake-like crystals filling the interparticle gaps. Comparative analysis indicated that the addition of GA or GO eliminated the cotton-like coating observed in AB/AP, instead promoting the growth of fine, fragmented structures interspersed among larger crystals. This observation suggested that the presence of GA and GO modified the AB/AP crystallization process, resulting in altered agglomeration morphologies. AB/AP(Δ) formed irregular bulk particles approximately 100 μm in size, containing abundant internal pores. AB/AP-GA(Δ) exhibited a morphology similar to APB, with comparable particle sizes around 100 μm. AB/AP-GO(Δ) similarly consisted of large, irregular particles with surface pores measuring about 100 μm. These surface pores originated from hydrogen evolution during the thermal treatment process, after which perchlorate ions combined with the foamed boron-nitrogen polymer to form the ammonia perchlorate borane derivative, APB.

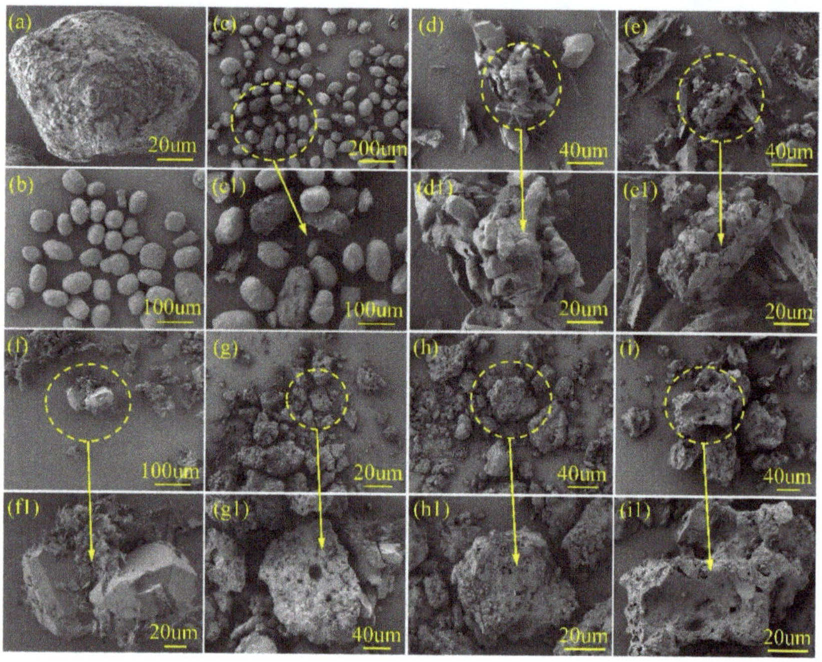

Figure 1. SEM images of AB/AP composites: (**a**) AB, (**b**) AP, (**c**) mechanical mixture AB/AP, (**d**) AB/AP, (**e**) AB/AP-GA, (**f**) AB/AP-GO, (**g**) AB/AP(Δ), (**h**) AB/AP-GA(Δ), and (**i**) AB/AP-GO(Δ).

Table 3 presents the elemental composition of the AB/AP composite. The elemental analysis results indicate that thermal treatment led to a reduction in the nitrogen (N) and hydrogen (H) content within the composite, which is attributed to both the release of hydrogen gas during the thermal treatment process and the generation of NH_3, N_2, and trace amounts of nitrogen oxides through AP's redox reaction. The escape of these gaseous products consequently caused a significant decrease in the N and H content in the calcined product.

Table 3. Elemental analysis test results.

Composites		N (%)	C (%)	H (%)	O (%)
Mechanical mixture AB/AP	Experimental	18.10	0	7.30	42.65
	Theoretical	18.87	0	6.74	43.13
AB/AP	Experimental	18.02	0	6.60	42.66
	Theoretical	18.87	0	6.74	43.13
AB/AP-GA	Experimental	17.83	1.54	6.46	42.34
	Theoretical	18.50	1.98	6.61	42.28
AB/AP-GO	Experimental	18.60	0.88	6.84	41.46
	Theoretical	18.62	0.80	6.65	43.09
AB/AP(Δ)	Experimental	10.54	0	3.30	43.21
	Theoretical	18.87	0	6.74	43.13
AB/AP-GA(Δ)	Experimental	13.12	1.63	4.23	42.35
	Theoretical	18.50	1.98	6.61	42.28
AB/AP-GO(Δ)	Experimental	13.36	0.82	3.95	41.63
	Theoretical	18.62	0.80	6.65	43.09

The oxygen (O) content of AB/AP-GO and AB/AP-GO(Δ) decreased significantly compared with the theoretical results. This is due to a large number of bubbles emerging from the sample with added GO during the preparation process. It is speculated that the poor compatibility between GO and AB in water may lead to the hydrolysis reaction of AB showing strong reducibility, and the oxidation–reduction reaction with a small amount of ClO_4^- may lead to the reduction of the measured O content of AB/AP-GO and AB/AP-GO(Δ).

2.2. Chemical Bond Structure and Energetic Properties

Simultaneously, the XRD diffraction patterns (Figure 2) revealed that the diffraction peak positions of both AB/AP composite particles and modified AB/AP composites were essentially identical to those of the prepared mechanical mixture, differing only in peak intensity. Table 4 presents the unit cell parameters of the crystalline samples. The data indicate that the crystal structures of the three AB/AP composites underwent significant changes: while both AB and AP exhibit an orthorhombic system, the composite particles adopted a monoclinic system. Although the three composites share the same crystal system, AB/AP-GO belongs to a different space group compared to the other two. These results confirmed that the characteristic diffraction peaks of AB and AP remained present in the composites modified with GA or GO, but the incorporation of graphene-based materials did not facilitate the formation of new crystal planes between AB and AP. Instead, all composite particles retained a mechanically mixed co-particle morphology. The XRD powder diffraction patterns further demonstrate that AB and AP did not form co-crystals but rather produced a mechanically mixed co-particle system.

Table 4 lists the unit cell parameters of the thermally treated products, showing that while they share the same crystal system, their space groups differ. Furthermore, the diffraction peak positions of the three thermally treated products showed no significant overall differences, with only minor variations in intensity, indicating their fundamental structural similarity.

The FTIR and Raman spectra of the composite were tested, and the results are shown in Figure 3. The characteristic peaks corresponding to each functional group are shown in Tables 5 and 6.

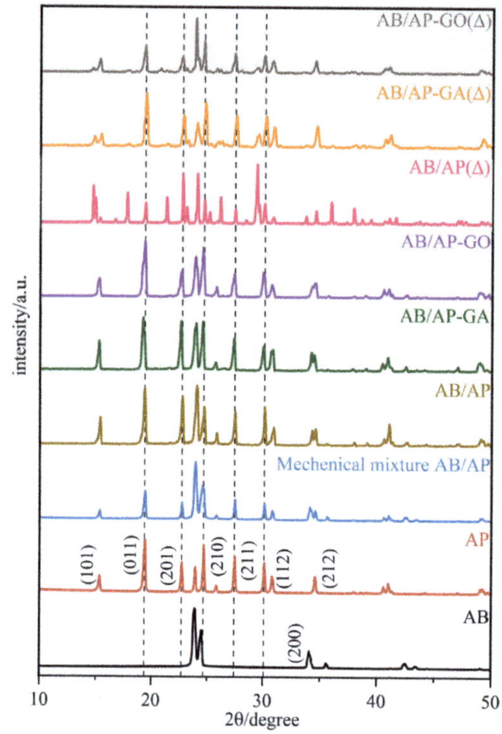

Figure 2. XRD pattern of composite particles.

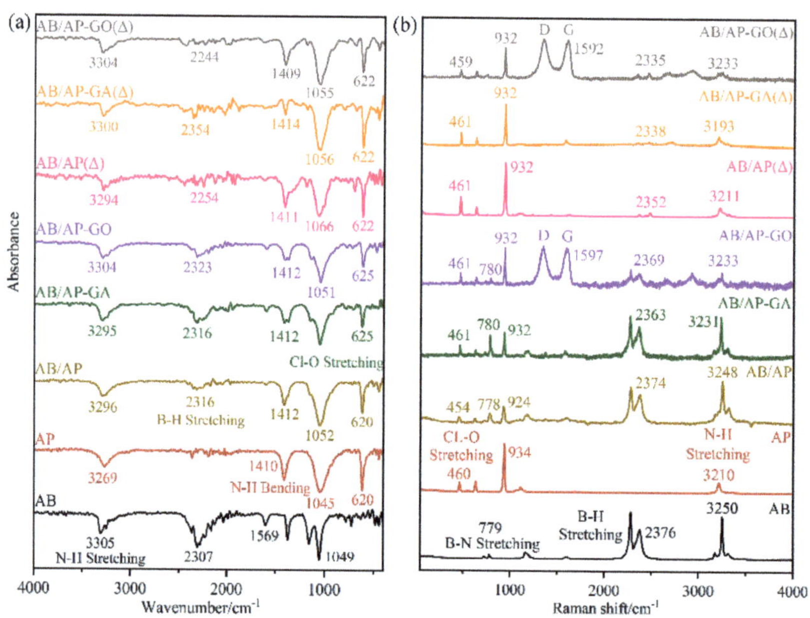

Figure 3. FTIR and Raman spectra of composite particles. (**a**) FTIR (**b**) Raman of Composite Particles.

Table 4. Unit cell parameters of composite particles.

	AB	AP	Mechanical Mixture AB/AP	AB/AP	AB/AP-GA	AB/AP-GO	AB/AP(Δ)	AB/AP-GA(Δ)	AB/AP-GO(Δ)
Crystall System	Orthorhombic	Orthorhombic	Orthorhombic	Monoclinic	Monoclinic	Monoclinic	Orthorhombic	Monoclinic	Monoclinic
Space Group	$Cmc2_1$	Pnma	Pccn	P2	P2	C2	Pccn	P2	$P2_1/C$
a/Å	12.668	9.226	11.716	6.906	11.516	8.956	15.896	6.773	14.003
b/Å	10.825	5.815	13.329	10.796	11.678	8.147	14.437	11.474	13.195
c/Å	8.183	7.456	9.850	6.062	8.417	10.021	11.936	19.230	6.637
α/deg	90.0	90.0	90.0	90.0	90.0	90.0	90.0	90.0	90.0
β/deg	90.0	90.0	90.0	123.5	137.6	113.5	90.0	99.2	104.0
γ/deg	90.0	90.0	90.0	90.0	90.0	90.0	90.0	90.0	90.0
Vol./Å^3	1122.2	400.0	1538.1	376.8	763.4	670.6	2739.2	1475.1	1190.0

Table 5. FTIR of AB, AP, and AB/AP composites.

cm^{-1}	AB	AP	AB/AP	AB/AP-GA	AB/AP-GO	AB/AP(Δ)	AB/AP-GA(Δ)	AB/AP-GO(Δ)
ClO_4^- Stretching	— —	620 1045	620 1052	625 1052	620 1051	622 1066	622 1056	622 1055
N-H Stretching	3305	3269	3296	3295	3304	3294	3300	3304
N-H Bending	1371 — 1569	— 1410 —	— 1412 1597	1371 1412 1597	1375 1412 1597	— 1411 —	— 1414 —	— 1409 —
B-H Stretching	2307	—	2316	2316	2323	2254	2354	2244
B-H Bending	1149 1049	— —	— 1148	— 1149	— 1146	— 1195	— 1191	— 1189

Table 6. Raman spectroscopy of AB, AP, and AB/AP composites.

cm^{-1}	AB	AP	AB/AP	AB/AP-GA	AB/AP-GO	AB/AP(Δ)	AB/AP-GA(Δ)	AB/AP-GO(Δ)
Cl-O Stretching	— — —	460 629 934	454 625 924	461 625 932	461 630 932	622 — 1066	622 — 1056	622 — 1055
N-H Stretching	3250	3210	3248	3231	3233	3294	3300	3304
B-H Stretching	2376	—	2374	2363	2369	2352	2338	2335
B-N Stretching	779	—	778	780	780	—	—	—
D(Disordered State)	—	—	—	—	1338	—	—	1338
G(Graphitic State)	—	—	—	—	1597	—	—	1592

The characteristic peaks of Cl-O and B-H were found in the three composite samples, which also proved that there were two kinds of substances, AB and AP, present. However, compared with the raw material, the positions of the three characteristic peaks have been shifted to a certain extent, mainly due to the obvious changes in the positions of the stretching vibration peaks of B-H and N-H. It is proven that the hydrogen bond between $-NH_3$ and ClO_4^- has been enhanced due to the interaction of N-H···O, which leads to the change of the stretching vibration peak of N-H. The shift of the B-H stretching vibration may be due to the effect of the H-H bond in AB or the formation of a B-H···O hydrogen bond. The characteristic peak shifts of AB/AP-GO and AB/AP-GA were the same, and the D peak and G peak of graphene oxide were found at 1338 cm^{-1} and 1597 cm^{-1}. The vibration mode of these three composite particles has been changed because of intermolecular interactions, such as hydrogen bonds, and the intensity and position of the absorption peak have also changed, and it is no longer caused by a single component intermolecular interaction. It was proved that the composite particles were not a simple mechanical mixture, and a new phase was formed, which was different from the mechanical mixture.

The main components of the three heat treatment products are related compounds of amino borane perchlorate. Among them, the stretching vibration peak intensity of ClO_4^- is particularly significant, and there are also relatively weak B-H stretching vibration peaks and bending vibration peaks. It is worth noting that the characteristic peak positions of these three products are shifted compared with AP, which is mainly due to the formation of APB. This process leads to significant changes in the original functional groups and their interactions, leading to the shift of the characteristic vibration peak positions of B-H and N-H. Compared with AB, the B-N stretching vibration peak of the three heat-treated products disappeared completely. Due to the low content of AB, only part of AP participated in the formation of APB during the heat-treatment process. During this process, some cycloborane and APB sublimate and precipitate, which reduces the proportion of amino borane in the system. At this stage, the concentration of cyclization products is very low, below the detection limits of both FTIR and Raman spectroscopy, and this leads to the disappearance of the B-N stretching vibration peak. The shift of the characteristic peak position is due to the change in the interaction between the original AB and AP molecules upon generation of APB. The Raman test results are consistent with those of the FTIR results, which proves that the three products contain APB after heat treatment. These derivatives change the interaction between molecules and form a new phase.

To measure the mass calorific value of the materials, the combustion heat of the composite particles was determined through ignition under a 3 MPa oxygen atmosphere. The density of each material was measured using a densitometer, and the corresponding volumetric calorific values were then calculated. The specific experimental data are presented in Table 7.

Table 7. Density and heat of combustion of composite particles.

Atmosphere	Materials	Mass Heat of Reaction ($kJ \cdot g^{-1}$)	Density ($g \cdot cm^{-3}$)	Volumetric Heat of Reaction ($kJ \cdot cm^{-3}$)
O_2	AB	42.54	0.7581	32.25
O_2	AP	3.64	1.9502	7.10
O_2	Mechanical mixture AB/AP	11.42	1.2693	14.50
O_2	AB/AP	10.44	1.6047	16.75
O_2	AB/AP-GA	10.18	1.5375	15.65
O_2	AB/AP-GO	10.03	1.5176	15.22
O_2	AB/AP(Δ)	9.67	1.6604	16.07
O_2	AB/AP-GA(Δ)	8.93	1.6021	14.31
O_2	AB/AP-GO(Δ)	7.94	1.5796	12.54

The energetic properties of the materials were evaluated by measuring the combustion heat of composite particles under a 3 MPa oxygen atmosphere using bomb calorimetry, while the material densities were determined with a densitometer to obtain volumetric heat values (specific data presented in Table 7). The results demonstrate that pure AB achieved the mass-specific heat value of 42.54 $kJ \cdot g^{-1}$ in oxygen, whereas pure AP exhibited a mass-specific heat value of 3.64 $kJ \cdot g^{-1}$.

As shown in Table 7, the theoretical mass combustion heat of the AB/AP mechanical mixture reached 11.42 $kJ \cdot g^{-1}$, while the combustion heats of the three composite particles were relatively close to this value, indicating that the actual molar ratios of AB to AP in the composites were consistent with the theoretical design. However, the unmodified AB/AP exhibited higher combustion heat, demonstrating that the incorporation of graphene-based materials reduced the energetic performance of the composites. The combustion heats of all three thermally treated products showed significant decreases compared to the raw mechanical mixture: AB/AP(Δ) decreased by 15.3%, AB/AP-GA(Δ) by 21.8%, and AB/AP-GO(Δ) by 30.5%. This reduction was primarily attributed to the sublimation of some ammonia borane derivatives [32–36], which led to decreased energetic performance of the system. Nevertheless, the remaining energy performance still maintained a relatively high level, suggesting that the formation of APB [32–36,40] partially compensated for the energy loss. Furthermore, the mass combustion heat data revealed that APB itself possessed relatively high combustion heat. This finding further confirmed that the addition of graphene-based materials negatively impacted the energetic properties of the composite system. The experimental results demonstrated that while graphene modification affected combustion performance, the as-prepared APB derivatives helped maintain considerable energy output in the composites.

The density data in Table 7 reveal that the measured densities of the composite materials were generally higher than their theoretical values, demonstrating that the molecular interactions between AB and AP components extended beyond simple mechanical mixing to include significant intermolecular forces that promoted tight binding. Notably, the unmodified composites exhibited higher densities compared to their modified counterparts, as the incorporation of low-density graphene-based materials increased molecular dispersion within the modified composites, thereby reducing their overall density. After thermal treatment, all materials showed substantial density increases relative to theoretical values: AB/AP(Δ) by 30.8%, AB/AP-GA(Δ) by 26.2%, and AB/AP-GO(Δ) by 24.4%. These enhancements confirmed that the formation of APB during thermal processing altered the intermolecular interactions, leading to significant modifications in product density. Furthermore, the inherently high density of APB itself may have contributed to the observed increases in system density. The thermal treatment significantly enhanced the density of AB/AP composites, effectively addressing the issue of low density inherent to pure AB materials.

2.3. *Thermal Behavior and Decomposition Kinetics*

2.3.1. Non-Isothermal Mass Loss and Heat Flow Properties

The thermal reactivity of the raw materials AB, AP, and the composite particles was investigated using a differential scanning calorimeter (DSC). The experimental conditions were maintained as follows: argon gas flow rate at 40 mL/min, heating rate of 10 $K \cdot min^{-1}$, and a temperature range from 50 to 500 °C. The resulting TG-DTG and DSC curves are presented in Figure 4, and the thermal decomposition parameters are summarized in Tables 8 and 9.

As shown in Figure 4, the thermal decomposition of AB occurred in two distinct stages: (i) Between approximately 112 and 130 °C, the first decomposition stage took place (NH_3-BH_3 → -$[NH_2$-$BH_2]_n$- + H_2), releasing one hydrogen molecule. AB melted between 97 and 120 °C, indicating that the first decomposition step occurred simultaneously with melting. (ii) In the temperature range of 130–200 °C, the second decomposition stage proceeded (-$[NH_2$-$BH_2]_n$- → -$[NH{=}BH]_n$- + H_2), releasing an additional hydrogen molecule and leaving a residue of 41%. The total heat released during AB decomposition was measured at 616.3 $J \cdot g^{-1}$, consistent with previously reported thermal stability data for AB [32–37]. Additionally, an endothermic peak at 242.2 °C was identified, corresponding to the phase transition of AP from orthorhombic to cubic crystal structure. AP decomposition also exhibited two stages: (i) The low-temperature decomposition stage occurred between 274 and 300 °C, with an associated heat release of 361 $J \cdot g^{-1}$. (ii) The high-temperature decomposition stage took place from 300 to 382.2 °C, releasing 631 J·g^{-1} of heat and resulting in a final residue of 11.5%.

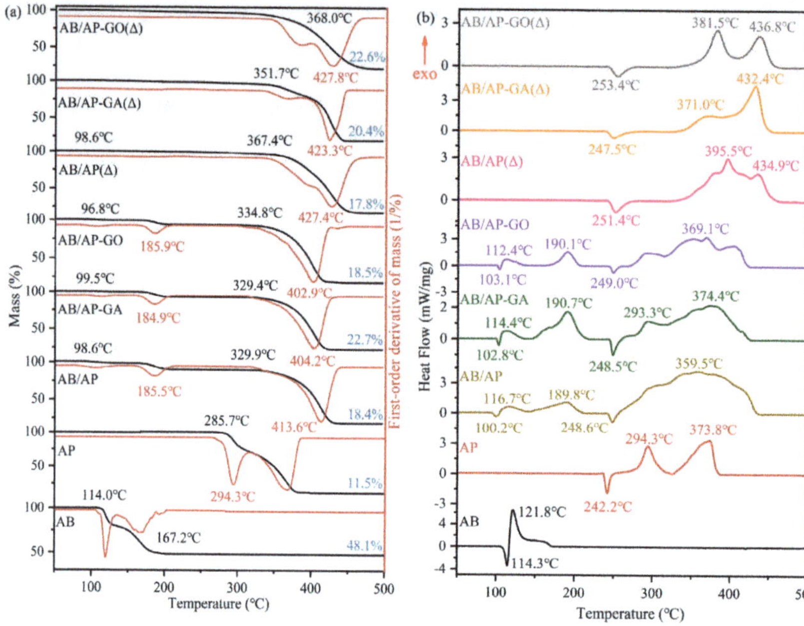

Figure 4. Thermal decomposition of AB, AP, and three composite particles: (**a**) TG-DTG curves and (**b**) DSC curves.

Table 8. The characteristic TG-DTG parameters of the thermal decomposition for AB, AP, and three composite particles.

Composites	TG Curves		DTG Curves			
	T_i (°C)	*ML* (%)	T_{p1} (°C)	T_{p2} (°C)	T_{p3} (°C)	T_{p4} (°C)
AB	114.0 149.3	21.8 30.1	119.7	167.2	—	—
AP	285.7 333.5	29.1 59.4	—	—	294.3	367.2
AB/AP	98.6 174.1 365.4	2.1 16.3 63.2	—	185.5	—	413.6
AB/AP-GA	99.5 174.8 363.2	0.7 6.5 70.1	—	184.9	—	404.2
AB/AP-GO	96.8 175.1 365.5	0.9 5.476.1	—	185.9	—	402.9
AB/AP(Δ)	367.4	82.2	—	—	—	427.4
AB/AP-GA(Δ)	351.7	79.6	—	—	—	423.3
AB/AP-GO(Δ)	368.0	77.4	—	—	—	427.8

Table 9. The characteristic DSC parameters of the thermal decomposition for AB, AP, and three composite particles.

Composites	DSC Curves							
	Endothermic Peaks				Exothermic Peaks			
	T_{p2} (°C)	ΔH_1 (J·g^{-1})	T_{p2} (°C)	ΔH_2 (J·g^{-1})	T_{p3} (°C)	ΔH_3 (J·g^{-1})	T_{p4} (°C)	ΔH_4 (J·g^{-1})
AB	114.3	88.5	—	—	121.5	616.3	—	—
AP	—	—	242.2	64.6	—	—	294.3 373.8	361.2 630.8
AB/AP	100.2	11.1	248.6	37.2	116.7 189.8	63.7 317.6	359.5	2996.6
AB/AP-GA	102.8	5.8	248.5	40.6	114.4 190.7	40.4 304.1	374.4	1184.2
AB/AP-GO	103.1	6.8	249.0	31.1	112.4 190.1	53.6 165.5	369.1	1777.2
AB/AP(Δ)	—	—	251.4	74.0	—	—	434.9	1098.3
AB/AP-GA(Δ)	—	—	247.5	67.6	—	—	432.4	1212.8
AB/AP-GO(Δ)	—	—	253.4	66.5	—	—	436.8	800.8

The AB/AP composite particles prepared by direct freeze-drying exhibited a two-stage decomposition profile, with each stage corresponding to the decomposition of AB and AP components, respectively. Notably, the low-temperature decomposition of AP disappeared, retaining only its high-temperature decomposition. The initial decomposition temperature of AP in the composite showed a significant increase of 70–80 °C compared to pristine AP. Furthermore, the heat release during AP decomposition in the composite (2997 J·g^{-1}) substantially exceeded that of pristine AP (962 J·g^{-1}). The residual mass ratio after decomposition of the AB/AP composite particles was measured at 18.4%. Similarly, the modified AB/AP composites only displayed high-temperature decomposition of AP with equivalent temperature elevation (70–80 °C). The decomposition heats for AB/AP-GA (1184 J·g^{-1}) and AB/AP-GO (1777 J·g^{-1}) composites both surpassed that of pristine AP (962 J·g^{-1}). The residual mass ratios after decomposition were determined to be 22.7% and 18.5% for AB/AP-GA and AB/AP-GO composites, respectively.

Comparative analysis revealed that all three types of prepared composite particles induced the appearance of two distinct exothermic peaks during AB decomposition, while eliminating the low-temperature decomposition of AP and retaining only its high-temperature decomposition with significantly elevated decomposition temperatures. This alteration in thermodynamic properties provided clear evidence that the prepared composite particles were not simple mechanical mixtures but rather demonstrated enhanced thermal stability due to the formation of intermolecular interactions. Furthermore, the experimental residue quantities of all three composites (18.4%, 22.7%, and 18.5%) showed close agreement with the theoretically calculated value of 18.8%. This consistency confirmed that the actual AB/AP ratios in the synthesized composites corresponded well with the designed experimental proportions.

Notably, the unmodified AB/AP composite particles exhibited significantly higher heat release compared to their modified counterparts, which aligned consistently with the combustion heat data. Pure AB typically left behind boron–nitrogen polymeric chains and borazine derivatives after decomposition at 200 °C, which retained certain hydrogen storage capabilities but required extremely high temperatures (2000 °C) for complete conversion to boron nitride. According to Refs. [32–38] and experimental phenomena, we found that the presence of AP fundamentally altered this decomposition pathway. The residual boron–nitrogen compounds participated in redox reactions with oxidative species generated during AP's high-temperature decomposition, thereby significantly enhancing the exothermic output. In contrast, while the modified composites (AB/AP-GA and AB/AP-GO) still demonstrated increased heat release, their energetic performance was compromised because: (1) the incorporated graphene materials were non-energetic themselves, and (2) they competed for reaction with the oxidative intermediates during AP decomposition. This competition resulted in incomplete oxidation of the boron–nitrogen species, ultimately leading to reduced heat release compared to the unmodified system.

In Figure 4, it can be observed that a portion of the substance had already decomposed before 200 °C, although the overall weight loss ratio remained extremely low (less than 1%). This phenomenon might have been caused by incomplete hyperchlorination of B-N polymers during the dehydrogenation process. Comparative analysis of the three thermal decomposition curves revealed that while the endothermic peak of AP persisted, no low-temperature decomposition of AP was detected in the AB/AP(Δ) sample. However, the characteristic two-stage decomposition of AP was clearly evident in both the AB/AP-GA(Δ) and AB/AP-GO(Δ) samples. Notably, all three heat-treated samples exhibited decomposition temperatures as high as 350 °C, approximately 250 °C higher than the decomposition temperature of the raw AB material, demonstrating the significantly enhanced thermal stability of the composite materials after heat treatment.

The decomposition heat release values of AB/AP(Δ) and AB/AP-GA(Δ) after heat treatment (1098 $J \cdot g^{-1}$ and 1213 $J \cdot g^{-1}$, respectively) were higher than that of pristine AP (962 $J \cdot g^{-1}$). In contrast, AB/AP-GO(Δ) showed a lower decomposition heat release (801 $J \cdot g^{-1}$) compared to AP. Notably, all three samples exhibited residual rates close to 18.8%, suggesting minimal loss of AB through sublimation during the heat treatment process.

Comparative analysis revealed that the thermal stability of all three heat-treated products was significantly enhanced relative to AP. Based on Refs. [32–38] and experimental phenomena, during heat treatment, only trace amounts of ammonia borane derivatives underwent sublimation. While AB/AP(Δ) showed no low-temperature AP decomposition, both AB/AP-GA(Δ) and AB/AP-GO(Δ) displayed this characteristic. This difference was attributed to the establishment of a new intermolecular interaction system between the APB and AP formed during heat treatment, which fundamentally altered the thermal stability of the products. However, the introduction of graphene-like materials weakened these newly formed intermolecular interactions, thereby allowing the low-temperature decomposition of AP to persist in both AB/AP-GA(Δ) and AB/AP-GO(Δ).

2.3.2. Decomposition Kinetic Parameters and Physical Models

Thermal analysis kinetics is widely employed in the field of energetic materials, enabling the investigation of reaction mechanisms and thermal stability through kinetic parameters [41–44]. To determine the thermal decomposition activation energy (E_a) of AB/AP composite particles, TG-DSC tests were performed at heating rates of 2.5, 5, 7.5, and 10 $K \cdot min^{-1}$. However, the experimental data obtained at 2.5 $K \cdot min^{-1}$ were unsuitable for calculating kinetic parameters. The DTG curves in Figure 4 reveal two distinct decomposition peaks at approximately 180 °C (Peak I) and 410 °C (Peak II) during AB/AP thermal decomposition. Additionally, calcinated AB/AP exhibited a prominent peak around 420 °C. Molecular dynamics calculations were subsequently conducted for these characteristic peaks using DTG data from different heating rates.

AB/AP systems were further investigated using the joint kinetic method [45] to calculate kinetic parameters and determine the most probable mechanism functions (Figure 5, Table 10). The analysis revealed that at Peak I, AB decomposition followed the F1 model (Unimolecular decay law), while AB/AP exhibited between F1 and D2 model (bidimensional particle shape). The high-temperature decomposition of AP consistently conformed from the A2 to the A3 model. At Peak II, AB/AP followed the L2 model (Random scission), AB/AP-GA and AB/AP-GO exhibited between the A2 and A3 model, confirming the disappearance of AP's low-temperature decomposition while retaining its high-temperature pathway. Notably, the calcinated products displayed distinct decomposition mechanisms: AB/AP(Δ), AB/AP-GA(Δ), and AB/AP-GO(Δ) all preferentially followed from the L2 to the R3 model (Phase boundary-controlled reaction). These findings indicated that: (1) AB/AP(Δ)'s decomposition mechanism differed significantly from raw AP, and that (2) the simultaneous decomposition of APB and AP under high-temperature conditions substantially modified the reaction mechanism.

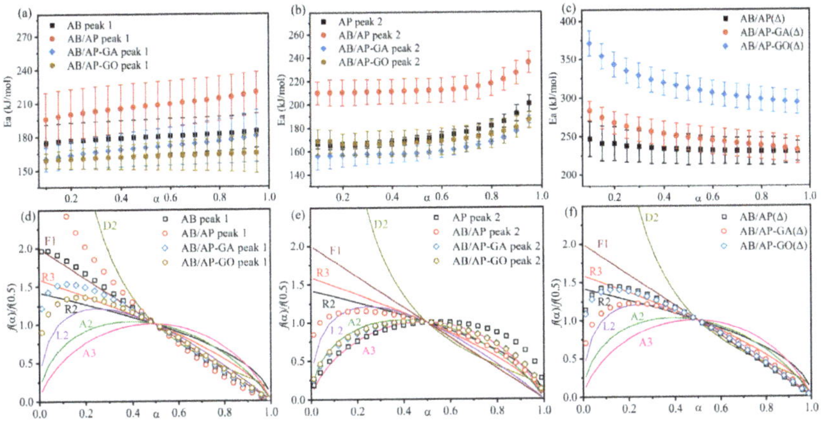

Figure 5. (**a**) The relationship between E_α and α during the thermal decomposition of AB and AB/AP composite particles (I) calculated by the Friedman method. (**b**) The relationship between E_α and α during the thermal decomposition of AB and AB/AP composite particles (II) calculated by the Friedman method. (**c**) The relationship between E_α and α during the thermal decomposition of AB/AP(Δ), AB/AP-GA(Δ), and AB/AP-GO(Δ) calculated by the Friedman method. (**d**) The thermal decomposition models of AP and AB/AP composite particles (I) calculated by the combined kinetic method. (**e**) The thermal decomposition models of AP and AB/AP composite particles (II) calculated by the combined kinetic method. (**f**) The thermal decomposition models of AB/AP(Δ), AB/AP-GA(Δ), and AB/AP-GO(Δ) calculated by the combined kinetic method.

Table 10. The decomposition kinetic parameters of the composite particles based on non-isothermal DTG curves.

Samples	Combined Kinetic Method				Friedman Method		Kissinger Method		
	m	n	$E_{a(1)}$/ kJ·mol^{-1}	cA/ min^{-1}	$E_{a(2)}$/ kJ·mol^{-1}	r	$E_{a(3)}$/ kJ·mol^{-1}	Log A	r
AB-I	0.029	1.150	187.0 ± 1.4	4.66×10^{21}	179.8	0.9925	61.7	10.51	0.9968
AP-II	0.521	0.424	171.5 ± 1.0	3.97×10^{12}	174.0	0.9994	41.9	6.71	0.9961
AB/AP-I	−0.123	1.251	218.8 ± 1.6	8.32×10^{23}	207.0	0.9894	58.3	9.86	0.9970
AB/AP-II	0.158	0.608	228.6 ± 3.1	1.14×10^{17}	214.2	0.9896	42.1	6.76	0.9980
AB/AP-GA-I	0.158	1.152	179.3 ± 1.4	2.24×10^{19}	169.8	0.9910	58.2	9.84	0.9972
AB/AP-GA-II	0.487	0.608	165.2 ± 0.9	1.90×10^{12}	162.1	0.9982	42.3	6.79	0.9973
AB/AP-GO-I	0.228	1.088	167.7 ± 1.3	9.33×10^{17}	162.1	0.9946	58.3	9.86	0.9967
AB/AP-GO-II	0.466	0.607	175.5 ± 2.3	1.77×10^{13}	170.2	0.9915	42.4	6.82	0.9974
AB/AP(Δ)	0.157	1.052	244.2 ± 2.5	7.01×10^{17}	235.7	0.9852	41.6	6.66	0.9982
AB/AP-GA(Δ)	0.269	0.953	253.6 ± 1.7	3.74×10^{18}	252.9	0.9958	41.5	6.65	0.9982
AB/AP-GO(Δ)	0.160	1.000	317.1 ± 1.6	1.47×10^{23}	321.8	0.9975	41.4	6.62	0.9983

The reactive molecular dynamic simulations on thermal decomposition confirmed that AP in the composite particles exclusively underwent high-temperature decomposition, validating the previously proposed decomposition mechanism for AB/AP composites. Based on the theoretical results, the thermal decomposition process revealed distinct stages: (1) AB initially decomposed, releasing hydrogen gas while generating boron–nitrogen polymeric chains and cyclic borazine derivatives. (2) These boron–nitrogen polymers subsequently encapsulated AP particles, altering the intermolecular interactions between AB and AP components. (3) With continued temperature increase, the modified molecular environment caused AP to bypass its characteristic low-temperature decomposition and proceed directly to high-temperature decomposition. Comparative analysis demonstrated significant mechanistic differences between AB/AP(Δ) and pure AP systems. The simultaneous decomposition of thermally generated APB and AP fundamentally modified the reaction pathway, resulting in a distinct decomposition mechanism for AB/AP(Δ).

2.4. Isothermal Pyrolysis Products and Decomposition Mechanisms

Pyrolysis GC-MS analysis was conducted to characterize the gaseous pyrolysis products of the three composite particle types. To simulate the heat treatment process, the pyrolysis chamber temperature was initially maintained at 100 °C with a heating rate of 20 °C·min^{-1}, yielding the TIC curves presented in Figure 6. Subsequently, the chamber temperature was elevated to 380 °C at the same heating rate for additional gas-phase product analysis, with corresponding TIC results shown in Figure 6. Since the predominant peaks in the TIC profiles appeared within the first 15 min, mass spectral analysis focused on the strong peak regions (TIC 1 and TIC 2) during this initial period. The time-resolved pyrolysis mass spectra are displayed in Figure S1, while the identified gaseous products are summarized in Table 11.

Table 11. The molecular formulas corresponding to the gaseous products from the pyrolysis mass spectrometry of the three composite particles at 100 °C and 380 °C.

m/z	Samples					
	AB/AP 100 °C	**AB/AP-GA 100 °C**	**AB/AP-GO 100 °C**	**AB/AP 380 °C**	**AB/AP-GA 380 °C**	**AB/AP-GO 380 °C**
14	N	N	—	—	—	—
17	—	—	—	NH_3	NH_3	NH_3
18	H_2O	H_2O	—	H_2O	H_2O	H_2O
28	N_2	N_2, CO	N_2, CO	N_2	N_2, CO	N_2, CO
30	—	—	—	NO	NO	NO
32	O_2	O_2	—	O_2	O_2	O_2
35	—	—	—	—	^{35}Cl	—
36	—	—	—	$H^{35}Cl$	$H^{35}Cl$	$H^{35}Cl$
38	—	—	—	—	$H^{37}Cl$	—
44	N_2O	N_2O, CO_2	N_2O, CO_2	N_2O	N_2O, CO_2	N_2O, CO_2
51–53	$(BH_xNH_y)_2$	$(BH_xNH_y)_2$	$(BH_xNH_y)_2$	$(BH_xNH_y)_2$	$(BH_xNH_y)_2$	$(BH_xNH_y)_2$
62	NO_3^-	NO_3^-	NO_3^-	NO_3^-	NO_3^-	NO_3^-
63	HNO_3	HNO_3	HNO_3	HNO_3	HNO_3	HNO_3
76–81			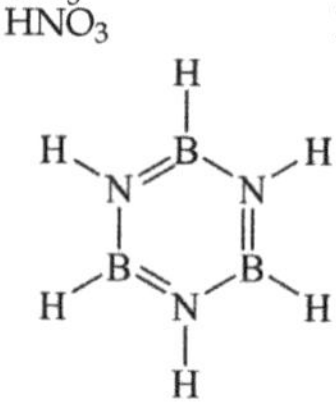			
100	$H^{35}ClO_4$	—	$H^{35}ClO_4$	—	—	—
101	$^{37}ClO_4^-$	—	—	—	—	—
102	$H^{37}ClO_4$	—	—	—	—	—
129–133	$NH_3BH_2{}^{37}ClO_4$	$NH_3BH_2{}^{37}ClO_4$	$NH_3BH_2{}^{37}ClO_4$	—	—	—
155–159	$NH_2BHNHBH$ $^{37}ClO_4^-$	$NH_2BHNHBH$ $^{37}ClO_4^-$	$NH_2BHNHBH$ $^{37}ClO_4^-$	—	—	—

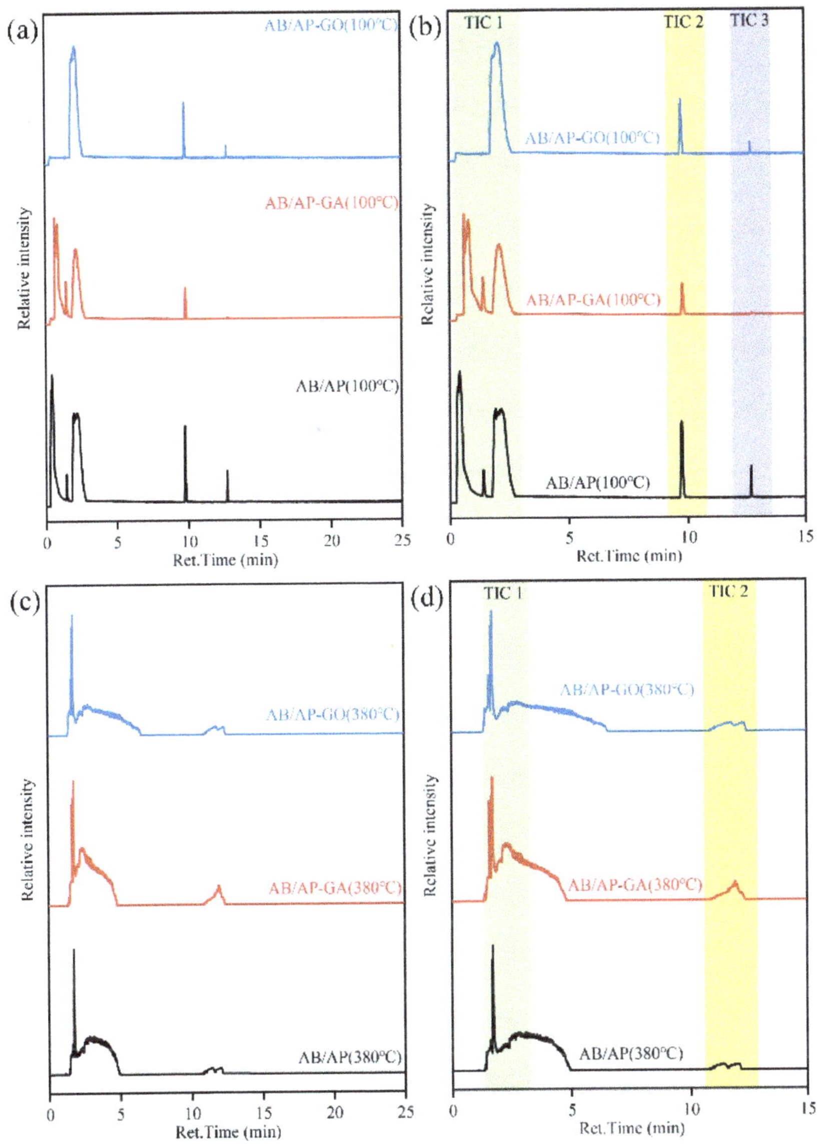

Figure 6. Total Ion Current (TIC) signal at a constant temperature of 100 °C and 380 °C: (**a**) 0–25 min at 100 °C, (**b**) 0–15 min at 100 °C, (**c**) 0–25 min at 380 °C, and (**d**) 0–15 min at 380 °C.

Comparative analysis of the TIC profiles revealed distinct differences in the strong peak positions among the three composite particles during 100 °C pyrolysis. Mass spectral characterization demonstrated that while peak positions varied, the gaseous products at these TIC strong peaks were essentially identical across all three materials. Specifically: N_2 and O_2 were detected in AB/AP and AB/AP-GA composites between 0.30 and 1.50 min. Cyclotriborazane derivatives appeared at 1.90–2.20 min. Modified composites exhibited N_2 and APB signatures at 9.70–9.80 min, while AB/AP showed perchloric acid and APB. During 12.60–12.80 min, N_2 and APB were consistently observed in modified composite products.

Comparing all gas products' mass spectra, it can be seen that at 100 °C, AP did not reach the decomposition temperature, and pure AP would not produce any gas products under heating conditions at 100 °C. However, H_2O was found in the gas-phase products N_2, O_2, $H^{35}Cl$, N_2O, and nitric acid [46] substances were also found, and the formation of cyclic compounds such as cyclotriborazane and APB was also discovered during the heat treatment process.

Comparative analysis of the TIC curve strong peak positions revealed that AB/AP-GO exhibited no detectable peaks between 0.30 and 1.50 min, indicating that graphene oxide incorporation significantly altered the decomposition pathway of AB/AP composites during heat treatment. Unlike AB/AP and AB/AP-GA, which sequentially released O_2 and N_2, AB/AP-GO directly co-generated multiple products at 1.90 min, including N_2, nitric acid [46], and cyclotriborazane derivatives [35–40]. These findings demonstrated that graphene oxide addition modified the reaction mechanism by suppressing AP's characteristic decomposition into O_2 and N_2 in the composite system.

During the heat treatment of AB/AP composites, AB decomposition produced H_2, cyclotriborazane derivatives, and boron–nitrogen polymeric species. The presence of AP induced significant chemical interactions, where the highly electronegative ClO_4^- ions promoted the formation of $NH_3BH_2^+$ species from AB, which subsequently combined with ClO_4^- to form APB [40]. Concurrently, boron–nitrogen polymers generated from AB decomposition also reacted with ClO_4^- to produce additional APB. This process caused partial AP to undergo redox reactions through ClO_4^- loss, while the remaining unreacted AP and newly formed APB collectively constituted the final heat-treated product.

Comparative analysis of the TIC profiles at 380 °C pyrolysis revealed nearly identical peak positions among the three composite particles. Mass spectral characterization confirmed similar gaseous product distributions at corresponding TIC peaks. Specifically: NH_3, N_2, N_2O, and cyclotriborazane derivatives were detected in all three composites between 1.58 and 1.59 min. Additional hypochlorous acid and nitric acid appeared at 1.70–1.75 min. During 2.30–2.85 min, NH_3, H_2O, and HCl were consistently identified in the decomposition products of all materials.

Comparative analysis of the mass spectra for all gaseous products revealed that the three composite materials generated essentially identical gas species during thermal cracking. Notably, the two modified AB/AP composites (AB/AP-GA and AB/AP-GO) produced additional CO and CO_2, demonstrating that oxidative species (particularly O_2) derived from AP decomposition oxidized the graphene-based additives during the process.

Gas-phase products N_2, O_2, $H^{35}Cl$, N_2O, and nitric acid [46] substances both originated from the gas products of the AP thermal cracking process [46]. Due to the stable presence of ^{35}Cl and ^{37}Cl in the natural chlorine element, the industrial grade ammonium perchlorate used may contain these two chlorine isotopes. The macromolecular compounds were $B_3N_3H_6$, which were gaseous products produced during the thermal decomposition of AB [32–36]. These substances still had a certain hydrogen storage capacity, but they would sublime into gas above 80 °C [34,40].

Comparative analysis of the TIC curves and TG-DSC data reveals that during heating, AB and AP in the AB/AP composite underwent separate decomposition processes. The thermal decomposition occurred in two distinct stages: Between 100 and 200 °C, AB decomposed to release hydrogen gas while generating boron–nitrogen polymeric chains and borazine derivatives, both of which maintained hydrogen storage capacity. These boron–nitrogen polymers subsequently encapsulated AP particles, significantly altering the AB–AP intermolecular interactions. Upon further heating to 350 °C, the modified molecular environment caused AP to bypass its characteristic low-temperature decomposition and proceed directly to high-temperature decomposition. This process generated oxidizing gases (H_2O, N_2, O_2, HCl) that reacted exothermically with the boron–nitrogen polymers, thereby enhancing the overall heat release during AP decomposition in the composite system.

3. Discussion

The thermal decomposition process of pure AB can be summarized from previous studies (see Figure 7): the initial stage of AB decomposition forms BH_2, BH, BH_3, etc. These compounds are subsequently converted into DADB, chain-like products, and H_2 [32–36]. Chainlike substances undergo self-catalytic reactions to form PABs. Under the catalysis of AB and NH_2BH_2, PAB would proliferate to form acyclic PABs or undergo cyclization to form BZ and PBZ. At the same time, the open-loop reaction of DADB gradually converts it into BZ, ultimately producing PBZ.

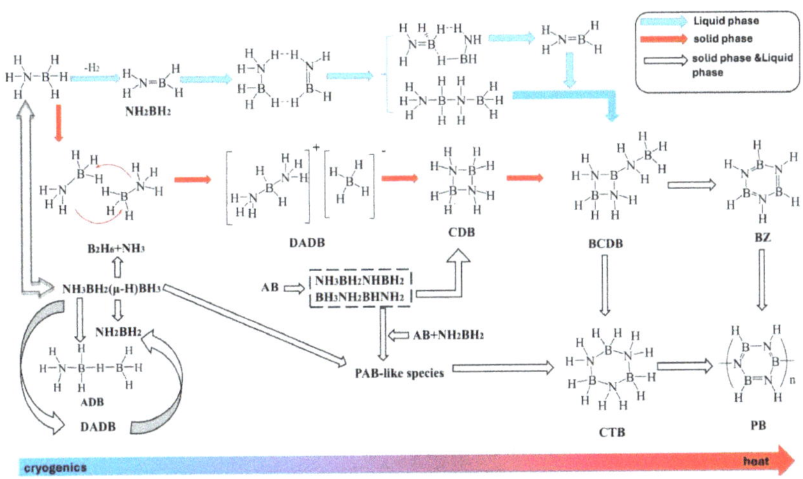

Figure 7. The mechanism of AB thermal decomposition.

Based on gas-phase product analysis from thermal cracking at 100 °C and 380 °C, the thermal decomposition process of AB/AP composite particles can be summarized (see Figure 8). During calcination at 100 °C, the binding between AP and AB initiates the following reactions: AB releases H_2 while forming borazine (BZ) and boron–nitrogen polymeric chains. The highly electronegative ClO_4^- from AP promotes $NH_3BH_2^+$ formation, which combines with ClO_4^- to yield APB [40]. Boron–nitrogen polymers and BZ from AB decomposition further react with ClO_4^- through dehydrogenation to produce additional APB. Concurrently, partial AP undergoes redox reactions via ClO_4^- loss, with the remaining AP and newly formed APB constituting the heat-treated product. Upon further heating above 350 °C, modified intermolecular interactions induce two parallel processes: direct high-temperature decomposition of AP, and oxidative decomposition of APB (major products: $4NH_3BH_2ClO_4 \rightarrow N_2\uparrow + 4HCl\uparrow + 2B_2O_3 + N_2O\uparrow + O_2\uparrow + 7H_2O\uparrow + H_2\uparrow$).

High-density AB/AP composite particles were successfully synthesized through direct freeze-drying at a 1:1 molar ratio (AB: AP). The formation of novel intermolecular interactions increased the composite density by 26.4% compared to mechanical mixtures, while simultaneously eliminating AP's low-temperature decomposition and raising its high-temperature decomposition onset by 80 °C. This method effectively resolved AB's inherent low-density limitation.

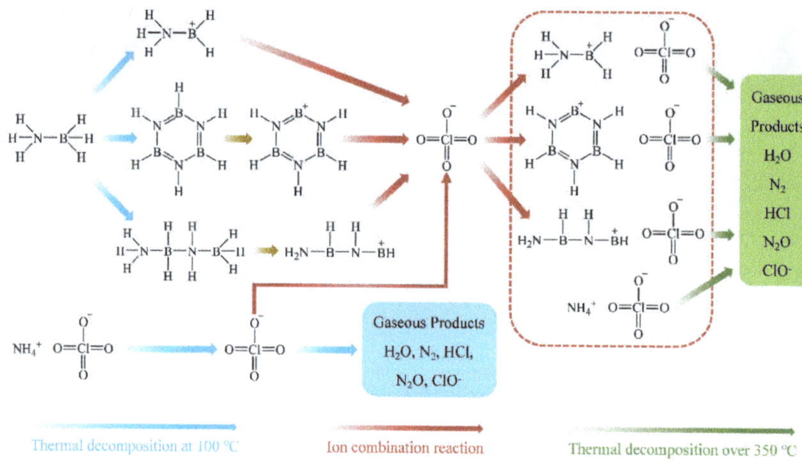

Figure 8. Thermal decomposition process of AB/AP composite particles.

Subsequent vacuum calcination of the composites produced AB/AP(Δ), which demonstrated superior thermal stability with a decomposition temperature 250 °C higher than pure AB and achieved a density of 1.6604 g/cm^3. APB in AB/AP(Δ) overcomes the limitations of low density and thermal instability of AB, and has the potential to be used as a high-energy-density additive for high-burning-rate solid propellant.

4. Materials and Methods

4.1. Materials

Ammonia borane (NH_3BH_3, AB) was purchased from Shanghai Aladdin Biochemical Technology Co., Ltd. Ammonium perchlorate (NH_4ClO_4, AP) was obtained from Henan Nayu Co., Ltd. Graphene (GA) was acquired from Shenzhen Hongdachang Technology Evolution Co., Ltd. Graphene oxide (GO) was procured from Nanjing Jicang Reagent Co., Ltd. Deionized water was sourced from Tianjin Fuyu Fine Chemical Co., Ltd.

4.2. Preparation of AB/AP Composites

4.2.1. Mechanical Mixtures of AB and AP

All AB/AP composites were prepared using the direct freeze-drying method. The preparation process is illustrated in Figure S2.

AB/AP: AB and AP were weighed in a molar ratio of 1:1 and dissolved in 20 mL of deionized water at room temperature. After thorough stirring to ensure complete dissolution, the solution was subjected to vacuum freeze-drying to obtain crystalline products. The yield was 99.7%.

In order to decrease the mechanical sensitivity of AB/AP composites, the carbon-modified AB/AP: 2 wt% of graphene (GA) or graphene oxide (GO) was dispersed in 20 mL of deionized water and ultrasonicated for 1 h. Subsequently, AB and AP were weighed in a molar ratio of 1:1 and dissolved in the ultrasonically dispersed solution at room temperature. After thorough stirring to ensure complete dissolution, the solution was vacuum freeze-dried to remove water and obtain crystalline products. The yield was 99.5% for both. The composite particles with added GA were designated as AB/AP-GA, and those with added GO were designated as AB/AP-GO.

AB/AP mechanical mixture: AB and AP were weighed in a molar ratio of 1:1 and placed in an agate mortar. A small amount of deionized water was added to moisten the mixture, which was then thoroughly ground. After the water evaporated, the mechanical mixture was obtained.

4.2.2. Calcinated Composites of AB and AP

The AB/AP, AB/AP-GA, and AB/AP-GO composites were placed in a vacuum drying oven and calcined at 100 °C for 2 h under vacuum to prepare high-density, thermally stable ammonium perchlorate borane (APB). The heat-treated AB/AP, AB/AP-GA, and AB/AP-GO were designated as AB/AP(Δ), AB/AP-GA(Δ), and AB/AP-GO(Δ), respectively. The preparation process is illustrated in Figure S3.

4.2.3. Characterizations and Theory

The morphology and microstructure of AB/AP composites and heat-treated AB/AP were characterized using scanning electron microscopy (SEM, Hitachi Regulus SU8230) at an acceleration voltage of 15.0 kV. The combustion heat was recorded by the ZDHW-HN9000A calorimeter. The combustion test was performed with 0.2 g of raw material in a 30 mL confined space, under an oxygen pressure of 3 MPa. The chemical bond structures of AB/AP composites and heat-treated AB/AP were investigated using Fourier transform infrared spectroscopy (FTIR, Brucker Corporation TENSOR II) and Raman spectroscopy (Thermal Fisher Scientific Dxr 2Xi). The thermal reactivity of AB/AP composites and calcinated AB/AP was evaluated by thermogravimetry-differential scanning calorimetry (TG-DSC) using a NETZSCH STA 449 simultaneous thermal analyzer under an argon atmosphere with a gas flow rate of 50 mL·min^{-1}. The composites were heated within a temperature range of 50–500 °C at heating rates of 2.5, 5, 7.5, and 10 K·min^{-1} to calculate their decomposition kinetics. The gaseous products from the thermal reaction of AB/AP composites and calcinated AB/AP were analyzed using pyrolysis-gas chromatography/mass spectrometry (Pyro-GC/MS, Frontier EGA/PY-3030D). The pyrolysis chamber temperature was set to 380 °C to study the pyrolysis products and to 100 °C to simulate the heat treatment environment, with a heating rate of 20 K·min^{-1}. The structures of AB/AP composites and heat-treated AB/AP were characterized using X-ray diffraction (XRD, Rigaku Ultima IV).

5. Conclusions

The AB/AP composite particles were prepared using a direct freeze-drying method, followed by thermal treatment to obtain calcinated AB/AP with enhanced thermal stability and density. The reaction mechanism of the thermal treatment process was investigated through pyrolysis gas-phase product analysis. The main findings were as follows:

High-density AB/AP composite particles with a 1:1 molar ratio were successfully prepared via direct freeze-drying. The establishment of novel intermolecular interactions between AB and AP increased the composite density by 26.4% compared to mechanical mixtures. The composite particles retained only AP's high-temperature decomposition, with an 80 °C higher initial decomposition temperature than pure AP, effectively addressing AB's low-density limitation.

Three types of composite particles were heated in a vacuum oven at 100 °C for 2 h, yielding oxygen/fuel-integrated APB with superior thermal stability. The thermal decomposition mechanism study revealed that: AB/AP initially decomposed during heat treatment, producing borazine derivatives and boron–nitrogen polymeric chains; AP presence facilitated $NH_3BH_2^+$ formation, which combined with ClO_4^- to generate APB; and partial AP underwent redox reactions through ClO_4^- loss, with the remaining AP and newly formed APB constituting AB/AP(Δ). Upon further heating to 350 °C, modified intermolecular interactions induced simultaneous AP high-temperature decomposition and

APB redox reactions, generating substantial gaseous products. However, carbon additives remained non-reactive during heat treatment, forming heterogeneous interfaces with APB and AP that weakened molecular interactions and reduced thermal stability in modified composites. APB demonstrated remarkable thermal stability (250 °C higher decomposition temperature than AB) and achieved a density of 1.6604 g/cm^3, overcoming AB's limitations of low density and poor thermal stability. These properties establish APB as a promising high-energy-density additive for high-burning-rate solid propellants.

Supplementary Materials: The following supporting information can be downloaded at: https://www.mdpi.com/article/10.3390/molecules30132680/s1, Figure S1: The mass spectra of pyrolysis of AB/AP composite particles and calcinated AB/AP at 100 °C and 380 °C; Figure S2: Synthesis of AB/AP composites; Figure S3: Synthesis of calcinated AB/AP crystals.

Author Contributions: Conceptualization, Y.Z. and Q.Y.; methodology, Y.Z.; formal analysis, Y.Z.; investigation, R.P.; resources, Q.Y.; data curation, Y.Z. and S.C.; writing-original draft preparation, Y.Z.; writing-review and editing, S.C. and Q.Y.; funding acquisition, S.C. and Q.Y. All authors have read and agreed to the published version of the manuscript.

Funding: This paper was funded by Basic Scientific Research Operating Funds for Central Universities (D5000240094).

Institutional Review Board Statement: Not applicable.

Informed Consent Statement: Informed consent was obtained from all subjects involved in the study.

Data Availability Statement: The data presented in this study are available upon request from the corresponding author.

Conflicts of Interest: The authors declare that they have no known competing financial interests or personal relationships that could have appeared to influence the work reported in this paper.

References

1. Maggi, F.; Gariani, G.; Galfetti, L.; DeLuca, L.T. Theoretical analysis of hydrides in solid and hybrid rocket propulsion. *Int. J. Hydrogen Energy* **2012**, *37*, 1760–1769. [CrossRef]
2. Yu, M.; Zhang, H.; Xie, W.; Li, Y.; Nie, H.; Yan, Q. Preparation and Reactivity Properties of Embedded-Coated AlH_3 Energetic Composite Particles. *Chin. J. Energ. Mater.* **2023**, *9*, 887–894. [CrossRef]
3. Kato, S.; Bielmann, M.; Ikeda, K.; Orimo, S.-I.; Borgschulte, A.; Züttel, A. Surface changes on AlH_3 during the hydrogen desorption. *Appl. Phys. Lett.* **2010**, *96*, 051912. [CrossRef]
4. Yu, M.-H.; Xu, R.; Xie, W.-X.; Li, Y.-J.; Nie, H.-Q.; Yan, Q.-L. High-Energy Composite Fuels with Improved Combustion Efficiency by Using AlH_3 Embedded with Al Particles. *ACS Appl. Mater. Interfaces* **2023**, *15*, 49611–49622. [CrossRef]
5. Yu, M.-H.; Yang, S.-L.; Xie, W.-X.; Li, Y.-J.; Li, H.-P.; Yan, Q.-L. Stabilization of AlH_3 Crystals by the Coating of Hydrophobic PFPE and Resulted Reactivity with Ammonium Perchlorate. *Langmuir* **2023**, *39*, 7863–7875. [CrossRef]
6. Ngene, P.; Berg, R.v.D.; Verkuijlen, M.H.W.; de Jong, K.P.; de Jongh, P.E. Reversibility of the hydrogen desorption from NaBH4 by confinement in nanoporous carbon. *Energy Environ. Sci.* **2011**, *4*, 4108–4115. [CrossRef]
7. Umegaki, T.; Xu, Q.; Kojima, Y. Porous Materials for Hydrolytic Dehydrogenation of Ammonia Borane. *Materials* **2015**, *8*, 4512–4534. [CrossRef]
8. Yu, M.-H.; Yang, S.-L.; Xie, W.-X.; Zhu, Z.-Y.; Li, H.-P.; Yan, Q.-L. Enhanced stability and combustion performance of AlH_3 in combination with commonly used oxidizers. *Fuel* **2022**, *331*, 125741. [CrossRef]
9. Yu, M.; Zhu, Z.; Li, H.-P.; Yan, Q.-L. Advanced preparation and processing techniques for high energy fuel AlH_3. *Chem. Eng. J.* **2021**, *421*, 129753. [CrossRef]
10. Staubitz, A.; Robertson, A.P.M.; Manners, I. Ammonia-Borane and Related Compounds as Dihydrogen Sources. *Chem. Rev.* **2010**, *110*, 4079–4124. [CrossRef]
11. Wang, P. Solid-state thermolysis of ammonia borane and related materials for high-capacity hydrogen storage. *Dalton Trans.* **2012**, *41*, 4296–4302. [CrossRef] [PubMed]
12. Moussa, G.; Moury, R.; Demirci, U.B.; Şener, T.; Miele, P. Boron-based hydrides for chemical hydrogen storage. *Int. J. Energy Res.* **2013**, *37*, 825–842. [CrossRef]

13. Shore, S.G.; Parry, R.W. Chemical Evidence for the Structure of the "Diammoniate of Diborane." II. The Preparation of Ammonia-Borane. *J. Am. Chem. Soc.* **1958**, *80*, 8–12. [CrossRef]
14. Figen, A.K.; Pişkin, M.B.; Coşkuner, B.; İmAmoğlu, V. Synthesis, structural characterization, and hydrolysis of Ammonia Borane (NH3BH3) as a hydrogen storage carrier. *Int. J. Hydrogen Energy* **2013**, *38*, 16215–16228. [CrossRef]
15. Boldyrev, V. Thermal decomposition of ammonium perchlorate. *Thermochim. Acta* **2006**, *443*, 1–36. [CrossRef]
16. Demirci, U.; Bernard, S.; Chiriac, R.; Toche, F.; Miele, P. Hydrogen release by thermolysis of ammonia borane NH3BH3 and then hydrolysis of its by-product [BNHx]. *J. Power Sources* **2011**, *196*, 279–286. [CrossRef]
17. Gutowska, A.; Li, L.; Shin, Y.; Wang, C.M.; Li, X.S.; Linehan, J.C.; Smith, R.S.; Kay, B.D.; Schmid, B.; Shaw, W.; et al. Nanoscaffold Mediates Hydrogen Release and the Reactivity of Ammonia Borane. *Angew. Chem. Int. Ed. Engl.* **2005**, *44*, 3578–3582. [CrossRef]
18. Neiner, D.; Karkamkar, A.; Linehan, J.C.; Arey, B.; Autrey, T.; Kauzlarich, S.M. Promotion of Hydrogen Release from Ammonia Borane with Mechanically Activated Hexagonal Boron Nitride. *J. Phys. Chem. C* **2008**, *113*, 1098–1103. [CrossRef]
19. Clark, T.J.; Lee, K.; Manners, I. Transition-Metal-Catalyzed Dehydrocoupling: A Convenient Route to Bonds between Main-Group Elements. *Chem. A Eur. J.* **2006**, *12*, 8634–8648. [CrossRef]
20. Langmi, H.W.; McGrady, G.S. Non-hydride systems of the main group elements as hydrogen storage materials. *Coord. Chem. Rev.* **2007**, *251*, 925–935. [CrossRef]
21. Baitalow, F.; Baumann, J.; Wolf, G.; Jaenicke-Rößler, K.; Leitner, G. Thermal decomposition of B–N–H compounds investigated by using combined thermoanalytical methods. *Thermochim. Acta* **2002**, *391*, 159–168. [CrossRef]
22. Weismiller, M.; Wang, S.; Chowdhury, A.; Thynell, S.; Yetter, R. Confined rapid thermolysis studies of ammonia borane. *Thermochim. Acta* **2013**, *551*, 110–117. [CrossRef]
23. Baitalow, F.; Wolf, G.; Grolier, J.-P.; Dan, F.; Randzio, S. Thermal decomposition of ammonia–borane under pressures up to 600bar. *Thermochim. Acta* **2006**, *445*, 121–125. [CrossRef]
24. Demirci, U.B. Ammonia borane, a material with exceptional properties for chemical hydrogen storage. *Int. J. Hydrogen Energy* **2017**, *42*, 9978–10013. [CrossRef]
25. Neiner, D.; Luedtke, A.; Karkamkar, A.; Shaw, W.; Wang, J.; Browning, N.D.; Autrey, T.; Kauzlarich, S.M. Decomposition Pathway of Ammonia Borane on the Surface of Nano-BN. *J. Phys. Chem. C* **2010**, *114*, 13935–13941. [CrossRef]
26. Patel, N.; Kale, A.; Miotello, A. Improved dehydrogenation of ammonia borane over Co-P-B coating on Ni: A single catalyst for both hydrolysis and thermolysis. *Appl. Catal. B Environ.* **2012**, *111–112*, 178–184. [CrossRef]
27. Kang, X.; Fang, Z.; Kong, L.; Cheng, H.; Yao, X.; Lu, G.; Wang, P. Ammonia Borane Destabilized by Lithium Hydride: An Advanced On-Board Hydrogen Storage Material. *Adv. Mater.* **2008**, *20*, 2756–2759. [CrossRef]
28. Nakagawa, T.; Burrell, A.K.; Del Sesto, R.E.; Janicke, M.T.; Nekimken, A.L.; Purdy, G.M.; Paik, B.; Zhong, R.-Q.; Semelsberger, T.A.; Davis, B.L. Physical, structural, and dehydrogenation properties of ammonia borane in ionic liquids. *RSC Adv.* **2014**, *4*, 21681–21687. [CrossRef]
29. Storozhenko, P.; Svitsyn, R.; Ketsko, V.; Buryak, A. Ammine borane: Synthesis And Physicochemical Characterization. *Russ. J. Inorg. Chem.* **2005**, *50*, 980–985.
30. Roy, B.; Hajari, A.; Kumar, V.; Manna, J.; Sharma, P. Kinetic model analysis and mechanistic correlation of ammonia borane thermolysis under dynamic heating conditions. *Int. J. Hydrogen Energy* **2018**, *43*, 10386–10395. [CrossRef]
31. Zhong, B.; Song, L.; Huang, X.X.; Xia, L.; Wen, G. First-principles investigation of ammonia borane for hydrogen storage. *Phys. Scr.* **2012**, *86*, 015606. [CrossRef]
32. Bowden, M.; Autrey, T.; Brown, I.; Ryan, M. The thermal decomposition of ammonia borane: A potential hydrogen storage material. *Curr. Appl. Phys.* **2008**, *8*, 498–500. [CrossRef]
33. Shaw, W.J.; Bowden, M.; Karkamkar, A.; Howard, C.J.; Heldebrant, D.J.; Hess, N.J.; Linehan, J.C.; Autrey, T. Characterization of a new phase of ammonia borane. *Energy Environ. Sci.* **2010**, *3*, 796–804. [CrossRef]
34. Fang, Z.; Luo, J.; Kang, X.; Xia, H.; Wang, S.; Wen, W.; Zhou, X.; Wang, P. Facile solid-phase synthesis of the diammoniate of diborane and its thermal decomposition behavior. *Phys. Chem. Chem. Phys.* **2011**, *13*, 7508–7513. [CrossRef]
35. Wang, J.; Freysoldt, C.; Du, Y.; Sun, L. First-Principles Study of Intrinsic Defects in Ammonia Borane. *J. Phys. Chem. C* **2017**, *121*, 22680–22689. [CrossRef]
36. Chen, X.-M.; Liu, S.-C.; Xu, C.-Q.; Jing, Y.; Wei, D.; Li, J.; Chen, X. Unravelling a general mechanism of converting ionic B/N complexes into neutral B/N analogues of alkanes: $H^{\delta+}\cdots H^{\delta-}$ dihydrogen bonding assisted dehydrogenation. *Chem. Commun.* **2019**, *55*, 12239–12242. [CrossRef]
37. Frueh, S.; Kellett, R.; Mallery, C.; Molter, T.; Willis, W.S.; King'oNdu, C.; Suib, S.L. Pyrolytic Decomposition of Ammonia Borane to Boron Nitride. *Inorg. Chem.* **2010**, *50*, 783–792. [CrossRef]
38. Baier, M.J.; Ramachandran, P.V.; Son, S.F. Characterization of the Hypergolic Ignition Delay of Ammonia Borane. *J. Propuls. Power* **2019**, *35*, 182–189. [CrossRef]
39. Biswas, P.; Wang, Y.; Herrera, S.; Ghildiyal, P.; Zachariah, M.R. Catalytic Cleavage of the Dative Bond of Ammonia Borane by Polymeric Carbonyl Groups for Enhanced Energy Generation. *Chem. Mater.* **2023**, *35*, 954–963. [CrossRef]

40. Biswas, P.; Ghildiyal, P.; Kwon, H.; Wang, H.; Alibay, Z.; Xu, F.; Wang, Y.; Wong, B.M.; Zachariah, M.R. Rerouting Pathways of Solid-State Ammonia Borane Energy Release. *J. Phys. Chem. C* **2021**, *126*, 48–57. [CrossRef]
41. Bélanger-Chabot, G.; Rahm, M.; Haiges, R.; Christe, K.O. Ammonia–(Dinitramido)boranes: High-Energy-Density Materials. *Angew. Chem. Int. Ed. Engl.* **2015**, *54*, 11730–11734. [CrossRef] [PubMed]
42. Kissinger, H.E. Reaction Kinetics in Differential Thermal Analysis. *Anal. Chem.* **1957**, *29*, 1702–1706. [CrossRef]
43. Vyazovkin, S.; Burnham, A.K.; Criado, J.M.; Pérez-Maqueda, L.A.; Popescu, C.; Sbirrazzuoli, N. ICTAC Kinetics Committee recommendations for performing kinetic computations on thermal analysis data. *Thermochim. Acta* **2011**, *520*, 1–19. [CrossRef]
44. Budrugeac, P. A simple and precise differential incremental isoconversional method to kinetic analysis of heterogeneous processes under arbitrary temperature programs. *Thermochim. Acta* **2018**, *661*, 116–123. [CrossRef]
45. Pérez-Maqueda, L.A.; Criado, J.M.; Sánchez-Jiménez, P.E. Combined Kinetic Analysis of Solid-State Reactions: A Powerful Tool for the Simultaneous Determination of Kinetic Parameters and the Kinetic Model without Previous Assumptions on the Reaction Mechanism. *J. Phys. Chem. A* **2006**, *110*, 12456–12462. [CrossRef]
46. Góbi, S.; Zhao, L.; Xu, B.; Ablikim, U.; Ahmed, M.; Kaiser, R.I. A vacuum ultraviolet photoionization study on the thermal decomposition of ammonium perchlorate. *Chem. Phys. Lett.* **2018**, *691*, 250–257. [CrossRef]

molecules

MDPI

Article

Dense Hydrogen-Bonded Assembly of Hydrogen-Rich Cations and Pentazolate Anions: A Series of Highly Insensitive Ionic Salts

Lianghe Sun, Hongwei Zhu, Shuaijie Jiang, Xiaofeng Yuan, Guoping Lu, Ming Lu * and Yuangang Xu *

School of Chemistry and Chemical Engineering, Nanjing University of Science and Technology, Nanjing 210094, China; 18112254480@163.com (L.S.); 18895545396@163.com (H.Z.); jsj2020@njust.edu.cn (S.J.); yuan_xiaof@163.com (X.Y.); glu@njust.edu.cn (G.L.)

* Correspondence: luming@njust.edu.cn (M.L.); yuangangxu@njust.edu.cn (Y.X.)

Abstract: Compounds containing the pentazolate anion (*cyclo*-N_5^-) represent a distinctive group of energetic materials that have received extensive attention in recent years. *Cyclo*-N_5^- was used as a polynitrogen anion for the syntheses of energetic salts through metathesis reactions. Propamidinium (**1**), 5-amino-4-carbamoyl-1*H*-imidazol-3-ium (**2**), (1*H*-1,2,3-triazol-4-yl)methanaminium (**3**), 5-amino-4*H*-1,2,4-triazol-1-ium (**4**), 5-amino-3-methyl-4*H*-1,2,4-triazol-1-ium (**5**), and amino(pyrimidin-2-yl)methaniminium (**6**) pentazolates were obtained with high yields (>80%), and their crystal structures were confirmed through single-crystal X-ray diffraction analyses. Hirshfeld surface analyses and 2D fingerprint plots generated by CrystalExplorer17 demonstrated that these compounds exhibited extensive hydrogen-bonding networks in their crystal packing. Mechanical sensitivity tests showed that all the prepared salts were highly insensitive (IS > 35 J, FS > 360 N), providing valuable insights for the further exploration of broader energetic materials containing *cyclo*-N_5^-.

Keywords: pentazolate anion; hydrogen bond; crystal structures; mechanical sensitivity

1. Introduction

Nitrogen-rich energetic materials represent a significant category of high-energy density materials that have significant applications in energetic systems, including propellant formulations, explosive compositions, and advanced energy storage technologies [1–5]. These nitrogen-rich compounds exhibited distinctive molecular architectures featuring a high density of N−N, C=N, and C−N bonds, which collectively enhance their positive heats of formation. The remarkable energy output characteristics of these materials are primarily derived from their substantial heats of formation, enabling the exothermic production of highly stable nitrogen gas (N_2) upon molecular decomposition. This characteristics position them as a highly promising category of eco-friendly energetic materials [6–8]. Building upon nitrogen-rich energetic materials, new investigations have revealed that converting them into ionic nitrogen-rich energetic salts can further enhance their performance. Compared to their neutral counterparts and conventional molecular energetic materials, nitrogen-rich energetic salts typically demonstrate superior physicochemical properties including reduced volatility, increased density, and enhanced thermal stability [9–11]. Among various nitrogen-rich structural motifs (such as pyrazole, tetrazole, and triazole derivatives), *cyclo*-pentazolate (*cyclo*-N_5^-) salts exhibit remarkable features, most notably their highly positive enthalpy of formation and environmentally friendly

https://doi.org/10.3390/molecules30122613

nitrogen gas being the primary detonation product. The molecular structure of *cyclo*-N_5^- salts contains metastable N-N and N=N bonds, which can release substantial dissociation energy upon conversion to the highly stable N≡N triple bonds. This distinctive property endows *cyclo*-N_5^- salts with broad application prospects in high-energy material fields, including propulsion systems, explosive charges, and pyrotechnic compositions [12,13].

However, in the development of *cyclo*-N_5^- salts, structural stabilization has always been the core challenge in design and experimental synthesis, as their stability directly determines the material's storage performance and application value. Although the *cyclo*-N_5^- anion possesses extremely high energy density, its metastable nature poses significant challenges for storage and practical applications [14,15]. Non-metallic cation-based *cyclo*-N_5^- salts may achieve kinetic stabilization through hydrogen bonding networks. In contrast with the typical neutral hydrogen bonds found in conventional CHON-based energetic materials (carbon (C), hydrogen (H), oxygen (O), and nitrogen (N)-based energetic compounds, exemplified by TNT, RDX, HMX, and CL-20), these salts form unique charged hydrogen bond systems despite the absence of covalent interactions between ions. These hydrogen bonds exist not only within intramolecular structural units but also extensively in intermolecular interactions, ultimately constructing complex multidimensional hydrogen bonding networks. Both theoretical and experimental studies have confirmed that hydrogen bonds play a crucial stabilizing role in ionic systems containing *cyclo*-N_5^- salt, endowing these materials with the advantage of low sensitivity [16,17].

To develop non-metallic *cyclo*-N_5^- salts with low sensitivities, we designed a novel series of compounds by utilizing the hydrogen-bond networks in ionic systems. We hypothesized that introducing a dense hydrogen-bond network would reduce the sensitivity of these *cyclo*-N_5^- salts while preserving their energetic performance. Both theoretical calculations and experimental characterizations consistently confirmed their low sensitivities and acceptable energetic properties, thereby validating our design strategy.

2. Results and Discussion

2.1. Design and Synthesis

Using a method developed by our group, silver pentazolate (AgN_5) was successfully synthesized [18]. All target salts were successfully synthesized by the metathesis reactions of AgN_5 with the corresponding hydrochloride salts [19], and all the yields exceeded 80%. Compounds **1–6** were synthesized as shown in Scheme 1. It was found that salts **1–6** were easily soluble in CH_3CH_2OH, CH_3OH, and H_2O. After evaporation at room temperature for 3–5 days, crystals were formed, and their molecular structures were uniquely determined by single-crystal X-ray diffraction.

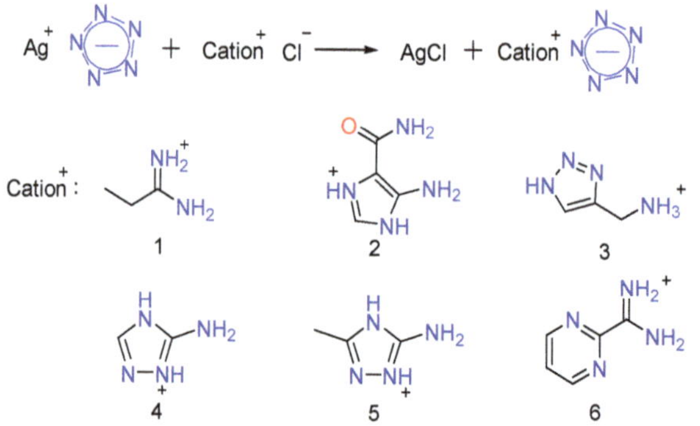

Scheme 1. Syntheses of *cyclo*-N_5^- salts **1–6**.

2.2. *Crystal Structures*

To gain deeper insights into the structural characteristics and intermolecular interactions of compounds **1–6**, single-crystal X-ray diffraction analyses were performed. Single crystals of compounds **2**, **4**, and **5** were obtained by slow evaporation from aqueous solution, while compounds **1**, **3**, and **6** were crystallized from methanol. The molecular structures of these compounds are presented in Figure 1.

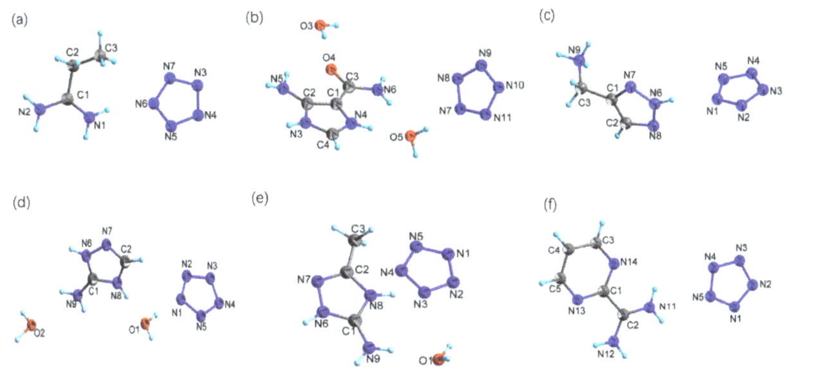

Figure 1. (**a–f**) Thermal ellipsoid diagrams (50% probability level) of compounds **1**, **2**·$2H_2O$, **3**, **4**·$2H_2O$, **5**·H_2O, and **6**.

Compound **1** crystallized in the monoclinic space group $P2_1/m$, with two molecules in the lattice units (Z = 2). As shown in Figure 2a, the asymmetric unit crystallography consists of one propamidinium cation and one *cyclo*-N_5^- anion. As shown in Figure 2a, each *cyclo*-N_5^- anion achieved stabilization via six distinct hydrogen bonds donated by four propamidinium cations (N1−H1A···N3 = 2.20 Å; N1−H1B···N6 = 2.13 Å; N2−H2A···N7 = 2.15 Å; N2−H2B···N4 = 2.14 Å; C3−H3B···N5 = 2.55 Å; C3−H3C···N6 = 2.61 Å). The N−N bond lengths in *cyclo*-N_5^- are 1.311(3), 1.319(3), 1.320(3), 1.316(3), and 1.320(3) Å, with an average N−N bond distance of 1.3172 Å. Compound **1** features layer-by-layer stacking (Figure 3a).

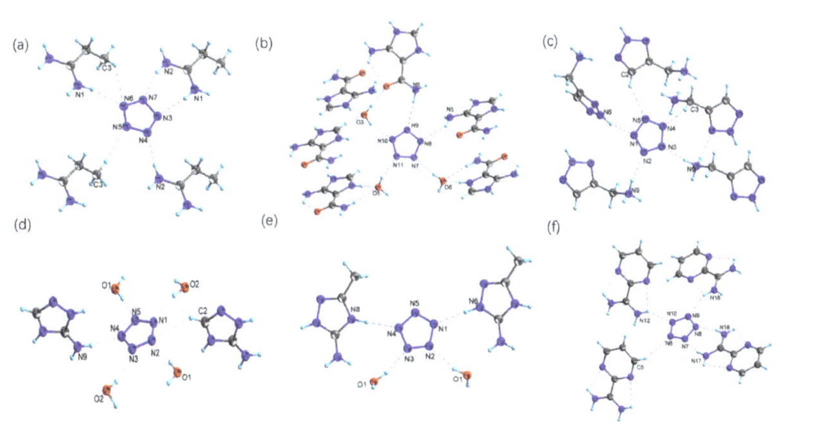

Figure 2. (**a–f**) Hydrogen bonds of the *cyclo*-N_5^- for compounds **1**, **2**·$2H_2O$, **3**, **4**·$2H_2O$, **5**·H_2O, and **6**. Dashed lines indicate strong hydrogen bonding.

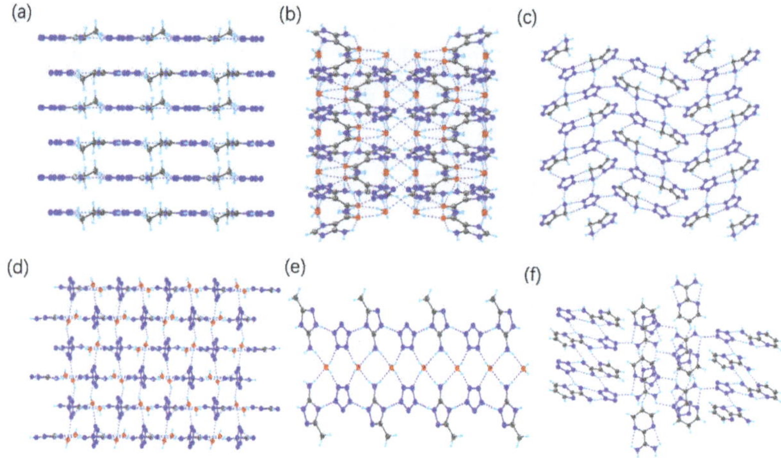

Figure 3. (**a–f**) The 3D structural layer network for compounds **1**, **2**·$2H_2O$, **3**, **4**·$2H_2O$, **5**·H_2O, and **6**. Dashed lines indicate strong hydrogen bonding.

Compound **2**·$2H_2O$ crystallized in the monoclinic space group *C*2/*c*, with eight molecules in the lattice units (Z = 8). As shown in Figure 2b, the asymmetric unit crystallography consists of one 5-amino-4-carbamoyl-1*H*-imidazol-3-ium cation, one *cyclo*-N_5^- anion, and two water molecules. The average N—N bond distance in the *cyclo*-N_5^- anion is 1.318 Å. As shown in Figure 2b, each *cyclo*-N_5^- anion achieved stabilization via five hydrogen bonds from two 5-amino-4-carbamoyl-1*H*-imidazol-3-ium cations (N8—H5A···N5 = 2.29(2) Å; N9—H6A···N6 = 2.22(2) Å) and three water molecules: (O3—H3A···N10 = 2.05(2) Å; O5—H5C···N7 = 1.99(2) Å; O5—H5D···N11 = 2.0.3(2) Å). The lengths of the hydrogen bonds are shorter than the sum of the van der Waals (vdW) radii (rw(N) + rw(N) = 3.20Å); therefore, a dense hydrogen bond network formed.

Both **3** and **4**·$2H_2O$ crystallized in the monoclinic space group $P2_1/c$, with four molecules in the lattice units (Z = 4) (Figure 2). The average N–N bond distances in **3** and **4**·$2H_2O$ are 1.315 Å and 1.3108 Å, respectively, which are slightly shorter than that in **2**. Each nitrogen atom of *cyclo*-N_5^- in **3** and **4**·$2H_2O$ serves as a hydrogen bond acceptor (Figure 3c,d); this results in the formation of dense 3D hydrogen-bonding networks.

Compound **5**·H_2O is a crystal belonging to the *Fdd*2 space group of the orthorhombic system, with sixteen molecules in the lattice units (Z = 16). The asymmetric unit crystallography consists of one 5-amino-3-methyl-4*H*-1,2,4-triazol-1-ium cation, one *cyclo*-N_5^- anion, and one water molecule (Figure 2e). The average N—N bond distance in the *cyclo*-N_5^- anion is 1.3178 Å. Each *cyclo*-N_5^- anion achieved stabilization via four hydrogen bonds from two 5-amino-3-methyl-4*H*-1,2,4-triazol-1-ium cations (N6—H6···N1 = 2.03(5) Å; N8—H8···N4 = 1.98(5) Å) and two water molecules (O1—H1A···N2 = 2.01(5) Å; O1—H1B···N3 = 1.93(7) Å). The N and C atoms of the 5-amino-3-methyl-4*H*-1,2,4-triazol-1-ium cations act only as hydrogen bond donors (Figure 3e). Unlike other compounds, in compound **5**·H_2O, only four nitrogen atoms in each *cyclo*-N_5^- unit act as hydrogen bond acceptors (Figure 3e).

Compound **6** crystallized in the triclinic space group *P*-1, with four molecules in the lattice units (Z = 4). As shown in Figure 2f, the asymmetric unit crystallography consists of one 2-amino(pyrimidin-2-yl)methaniminium cation and one *cyclo*-N_5^- anion. As shown in Figure 2f, each *cyclo*-N_5^- anion is stabilized through five hydrogen bonds formed by four amino(pyrimidin-2-yl)methaniminium cations and its five nitrogen atoms (C5—H5···N6 = 2.59 Å; N17—H17A···N7 = 2.07 Å; N18—H18A···N8 = 2.06 Å;

N18–H18B···N9 = 2.11 Å; N12–H12B···N10 = 2.29 Å). The five nitrogen atoms of *cyclo*-N_5^- in **6** are not perfectly coplanar, as evidenced by the largest torsion angle of 0.9° (N4–N3–N2–N1). The average N–N bond distance in the *cyclo*-N_5^- anion is 1.3146 Å.

In compounds **1**, **2**·$2H_2O$, **3**, **4**·$2H_2O$, **5**·H_2O, and **6**, the average N–N bond lengths in *cyclo*-N_5^- are 1.3172, 1.318, 1.315, 1.3108, 1.3178, and 1.3146 Å. The average N–N bond distance is shorter than that of most reported *cyclo*-N_5^- salts, indicating aromatic characteristics, which are consistent with stable *cyclo*-N_5^- salts [20–23]. Experimental data reveal that the hydrogen bond distances in this system range from 1.93 to 2.61 Å (Tables S6, S10, S14, S18, S22 and S26), which are significantly shorter than those in conventional *cyclo*-N_5^- salts [19], suggesting stronger hydrogen-bonding interactions. Salts **1**, **2**·$2H_2O$, **3**, **4**·$2H_2O$, **5**·H_2O, and **6** exhibit dense hydrogen-bonding interactions between the *cyclo*-N_5^- anions and adjacent cations, as well as water molecules, as clearly shown in Figure 3. This result is further supported by Hirshfeld surface analysis, where a distinct spike is observed in the lower-left corner of the 2D fingerprint plot (N···H interactions > 77%)

2.3. Vibrational Spectroscopy

The infrared (IR) spectra of compounds **1–6** were identified as providing valuable insights into intramolecular and intermolecular vibrational bonds. From the IR spectra of the six salts (Figure 4), it can be observed that the infrared vibrational absorption of *cyclo*-N_5^- lies between 1211 and 1229 cm^{-1}, which shows excellent agreement with the frequencies reported for *cyclo*-N_5^- salts (NH_4N_5, $N_2H_5N_5$, and $[Na(H_2O)(N_5)]·2H_2O$) [19,20]. The stretching vibrations of the N–H bonds in **1–6** were observed between 3500 and 3000 cm^{-1}.

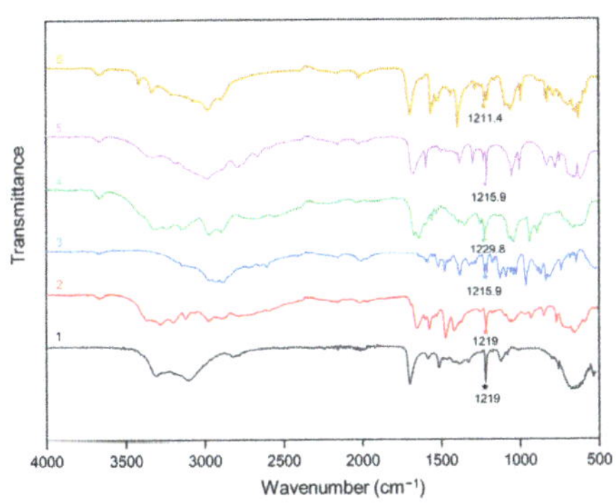

Figure 4. IR spectra of compounds **1–6**.

2.4. Thermal Stability

The thermal stabilities of compounds **1–6** were investigated using differential scanning calorimetry (DSC). As shown in Figure 5, compounds **1–6** all decomposed (**1**:121 °C; **2**:107 °C; **3**:118 °C; **4**:120 °C; **5**:108 °C; **6**:117 °C) without melting in the range of 50–300 °C. The thermal analyses in this study demonstrate that dense hydrogen-bonding networks enhance the thermal stabilities of *cyclo*-N_5^- salts. Salts **1–6** exhibit higher decomposition temperatures ($T_{peak} \geq 107$ °C) compared to $N(CH_3)_4N_5$, $[DAG]N_5$, and $[EDA](N_5)_2$ [14,19,22].

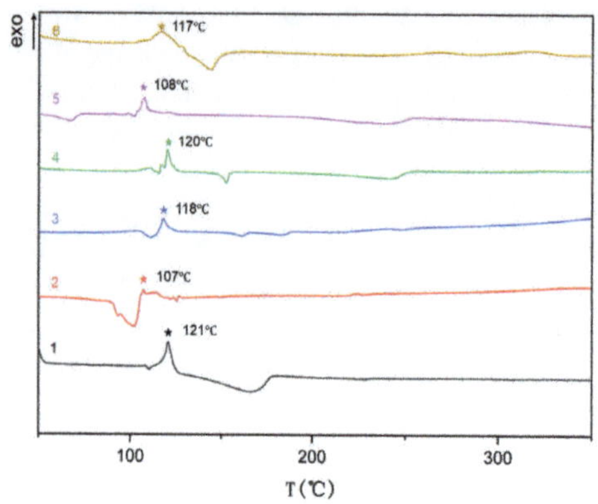

Figure 5. DSC curves for **1–6** at a heating rate of 5 °C min^{-1}.

2.5. Weak Interactions

To understand the interactions involving *cyclo*-N_5^-, the CrystalExplorer17 [24] program was employed to generate and analyze the two-dimensional (2D) fingerprint plots and Hirshfeld surfaces of the crystals. These results were utilized to assess the weak interactions in *cyclo*-N_5^- salts, where the red and blue regions correspond to high and low close-contact densities, respectively [25].

For **1**, **2**·$2H_2O$, **3**, **4**·$2H_2O$, **5**·H_2O, and **6**, a prominent peak at the bottom left (N···H interactions) can be seen in the 2D fingerprint plots (Figure 6). These N···H contacts contribute more than 77.3% of the total interactions (Figure 7), indicating that hydrogen bonding plays a dominant role in driving the formation of *cyclo*-N_5^--based crystals. The significant contribution of hydrogen bonds not only stabilizes the crystal structure but also facilitates the arrangement of *cyclo*-N_5^- anions and cations into a well-defined layered or 3D network. This highlights the critical importance of hydrogen bonding in the design and synthesis of *cyclo*-N_5^--based energetic materials with enhanced stability and performance.

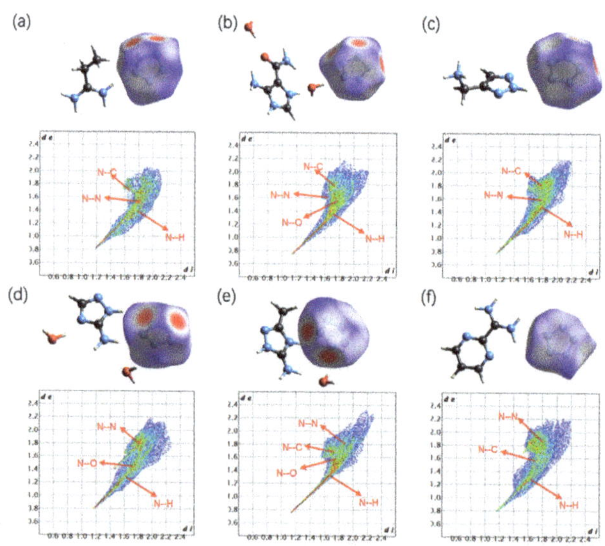

Figure 6. (**a–f**) Hirshfeld surfaces and 2D fingerprint plots of *cyclo*-N_5^- in compounds **1**, **2**·$2H_2O$, **3**, **4**·$2H_2O$, **5**·H_2O, and **6**.

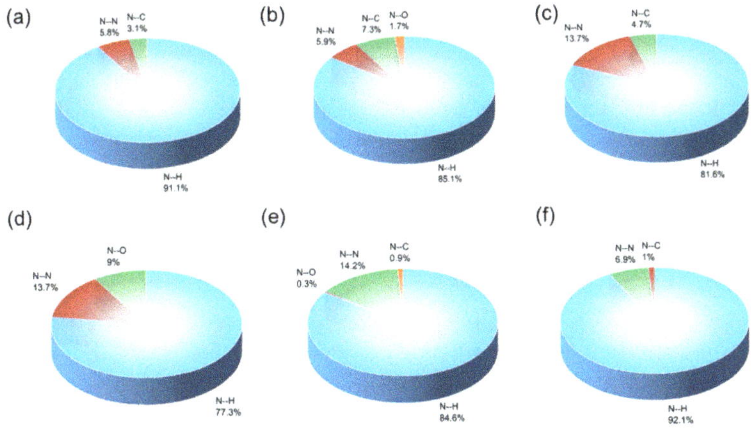

Figure 7. (**a–f**) Individual atomic contact percentage contributions of *cyclo*-N_5^- in compounds **1**, **2**·$2H_2O$, **3**, **4**·$2H_2O$, **5**·H_2O, and **6**.

2.6. Physical Chemistry and Energetic Properties

The solid-phase heats of formation for anhydrous **1–6** were calculated using the gas phase heats of formation and heats of phase transition (lattice energy) based on Hess's law of constant summation (Born–Haber energy cycle; see the Supplementary Material) [19,26,27]. The heats of formation of the *cyclo*-N_5^- anion and the corresponding cations were calculated using the Gaussian 09 (Revision A.02) suite of programs [28–30]. Salts **3** and **4** exhibit relatively high heats of formation, reaching 533.86 kJ mol^{-1} and 440.9 kJ mol^{-1}, respectively (Table 1).

Table 1. Physicochemical and energetic properties of compounds **1–6**.

Comp.	ρ [a] (g cm^{-3})	T_d [b] (°C)	IS [c] (J)	FS [d] (N)	ΔH_f [e] (kJ mol^{-1})	D [f] (m s^{-1})	P [g] (GPa)
1	1.307	121	>40	>360	213.97	7122	14.6
2	1.575	107	>40	>360	210.66	7791	23.3
3	1.541	118	>35	>360	533.86	8288	22.6
4	1.580	120	>35	>360	440.90	8162	25.1
5	1.430	108	>40	>360	402.99	7561	20.0
6	1.485	117	>40	>360	376.29	8094	21.0
$C_2H_6N_{10}$ [19]	1.583	107	>40	>360	639.7	7824	24.5
NH_4N_5 [19]	1.486	106.5	8	130	269.1	7757	23.2
$N(CH_3)_4N_5$ [14]	1.245	81.6	35	>360	296.1	6300	14.0
[DAG]N_5 [19]	1.438	100.5	25	240	508.0	7505	21.2
[EDA]$(N_5)_2$ [22]	1.395	98.5	25	240	556.1	6896	17.5

[a] Density, measured using a gas pycnometer at 25 °C. [b] Decomposition temperature (peak). [c] Impact sensitivity (IS). [d] Friction sensitivity (FS). [e] Heat of formation. [f] Calculated detonation velocity. [g] Calculated detonation pressure. DAG, 1,3-diaminoguanidinium. EDA, monoethane-1,2-diaminium.

The detonation properties of compounds **1–6** were theoretically estimated based on experimental densities and calculated heats of formation using the EXPLO5 program (version 6.05.04) [31]. The calculated detonation velocity (*D*) and pressure (*P*) values fall in the range of 7122–8288 m s^{-1} and 14.6–25.1 GPa (Table 1). The detonation velocities of compounds **3** and **4** are higher than that of 3,4-diamino-1,2,4-triazolium pentazolate, $N(CH_3)_4N_5$, and NH_4N_5 [14,19].

2.7. Mechanical Sensitivity

The sensitivity is a key indicator for evaluating the safety and practical usability of energetic materials. To analyze the sensitivity properties of anhydrous **1–6**, both the impact sensitivity (IS) and friction sensitivity (FS) were measured using the standardized BAM test procedures [32]. Remarkably, all synthesized salts demonstrated exceptionally low sensitivities, with impact sensitivity values exceeding 35 J and friction sensitivity values surpassing 360 N (Table 1). These results demonstrate that the materials are highly resistant to inadvertent initiation, rendering them suitable for operational and long-term storage under practical conditions.

Low sensitivities can be attributed to the synergistic effect of a highly ordered crystalline packing structure and intermolecular hydrogen bonding interactions. Specifically, the 3D network structure constructed through dense hydrogen bonds can significantly enhance the lattice energy of the crystal, thereby greatly improving its structural stability. As shown in Table 1, salts **1–6** exhibit lower mechanical sensitivities than $N(CH_3)_4N_5$, $[DAG]N_5$, and $[EDA](N_5)_2$ [14,19,22]. Such properties are highly desirable for the development of next-generation energetic materials with enhanced safety profiles.

3. Materials and Methods

During experimental operations, extra care must be taken when handling these high-energy compounds, and masks, gloves, and goggles must be worn throughout the entire process. The necessary protective equipment must be provided and close at hand.

3.1. Reagents and Instruments

All chemicals employed in this study were analytical-grade materials procured from domestic manufacturers in China and used without further purification. ^{1}H NMR (500 MHz) and ^{13}C NMR (125.72 MHz) spectra were recorded using a Bruker AVANCE III 500 spectrometer (Germany). Differential Scanning Calorimetry (DSC) studies were carried out using a NETZSCH DSC 204 F1 Phoenix instrument (Germany) operated in a nitrogen atmosphere with a heating rate of 5 °C min^{-1}. IR spectra were performed with a Bruker ALPHA II Fourier transform infrared spectrometer (Germany). The single-crystal X-ray diffraction measurements for **1–6** were conducted using a Bruker Smart Apex II diffractometer (Germany) using Mo-Kα radiation (λ = 0.71073 Å) or Cu-Kα radiation (λ = 1.54178 Å) with a graphite monochromator at temperatures of 150, 170, 173, and 296 K. An Anton Paar Ultrapyc 5000 gas pyrometer (Austria) operating at 25 °C was employed to determine the densities of the samples. Impact sensitivity was tested with a BAM Fall-hammer (Czech), while friction sensitivity was assessed using a BAM friction tester (Czech).

3.2. Experimental Methods

Freshly synthesized AgN_5 (127 mg, 0.71 mmol) was introduced into an aqueous solution (15 mL deionized water) containing the chloride salt (0.428 mmol) under continuous stirring. After maintaining the reaction at room temperature for 1 h with agitation, the resulting silver chloride precipitate was separated by filtration. Subsequently, the filtrate was concentrated under reduced pressure to obtain the target compound.

Propamidinium pentazolate (**1**): Yield: 52.196 mg, 85 %. T_d: 118 °C. ^{1}H NMR (500 MHz, DMSO-d_6): δ = 9.09, 8.79, 2.39, 1.16 ppm. ^{13}C NMR (125.72 MHz, DMSO-d_6): δ = 172.56, 25.87, 11.51 ppm. IR: ν = 3106, 2823, 1702, 1514, 1219, 960 cm^{-1}.

5-Amino-4-carbamoyl-1*H*-imidazol-3-ium pentazolate (**2**): Yield: 70.384 mg, 83 %. T_d: 106 °C. ^{1}H NMR (500 MHz, DMSO-d_6): δ = 14.32, 8.59, 7.41, 6.45 ppm. ^{13}C NMR

(125.72 MHz, DMSO-d_6): δ = 161.58, 143.57, 128.87, 102.95 ppm. IR: ν = 3286, 2024, 1581, 1476, 1219, 1065 cm^{-1}.

(1*H*-1,2,3-triazol-4-yl)methanaminium pentazolate (**3**): Yield: 63.788 mg, 88 %. T_d: 116 °C. ^1H NMR (500 MHz, DMSO-d_6): δ = 11.95, 8.67, 8.01, 4.12 ppm. ^{13}C NMR (125.72 MHz, DMSO-d_6): δ = 140.13, 49.01, 34.13 ppm. IR: ν = 2932, 2015, 1481, 1383, 1215, 966 cm^{-1}.

5-Amino-4*H*-1,2,4-triazol-1-ium pentazolate (**4**): Yield: 54.127 mg, 81 %. T_d: 104 °C. ^1H NMR (500 MHz, DMSO-d_6): δ = 13.68, 8.29, 8.03 ppm. ^{13}C NMR (125.72 MHz, DMSO-d_6): δ = 151.17, 139.39 ppm. IR: ν = 3144, 2047, 1649, 1354, 1229, 944 cm^{-1}.

5-Amino-3-methyl-4*H*-1,2,4-triazol-1-ium pentazolate (**5**): Yield: 66.211 mg, 91 %. T_d: 108 °C. ^1H NMR (500 MHz, DMSO-d_6): δ = 7.87, 5.18, 3.60 ppm (s, 2 H). ^{13}C NMR (125.72 MHz, DMSO-d_6): δ = 151.55, 147.97, 11.37 ppm. IR: ν = 2987, 2160, 1689, 1386, 1215, 1058 cm^{-1}.

Amino(pyrimidin-2-yl)methaniminium pentazolate (**6**): Yield: 69.770 mg, 81%. T_d: 109 °C. ^1H NMR (500 MHz, DMSO-d_6): δ = 9.54, 9.15, 7.92 ppm. ^{13}C NMR (125.72 MHz, DMSO-d_6): δ = 160.54, 158.82, 153.33, 125.38 ppm. IR: ν = 2987, 2027, 1572, 1399, 1211, 1066 cm^{-1}.

3.3. Theoretical Methods

The elementary geometric optimization and the frequency analysis were performed at the level of the Becke three parameter, Lee-Yan-Parr (B3LYP) [33] functional with the 6-311++G** basis set. All of the optimized structures were characterized to be local energy minima on the potential surface without any imaginary frequencies. Atomization energies were calculated by the CBS-4M [34].

For energetic salts, the solid-phase heats of formation are calculated on the basis of a Born-Haber energy cycle (Scheme 2).

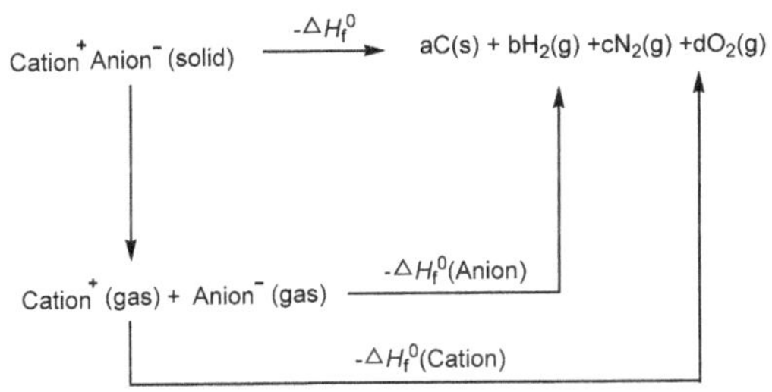

Scheme 2. Born-Haber cycle for the formation of energetic salts.

Based on a Born-Haber energy cycle, the heat of formation of a salt can be simplified by the formula given in Equation (1):

$$\Delta H_f \text{ (salt, 298 K)} = \Delta H_f \text{ (cation, 298 K)} + \Delta H_f \text{ (anion, 298 K)} - \Delta H_L \quad (1)$$

where ΔH_L is the lattice energy of the salts, which could be predicted by using the formula suggested by Jenkins et al. [35].

$$\Delta H_L = U_{POT} + [p(n_M/2 - 2) + q(n_X/2 - 2)]RT \quad (2)$$

where n_M and n_X depend on the nature of the ions, M_p^+ and X_q^-, and are equal to 3 for monatomic ions, 5 for linear polyatomic ions, and 6 for nonlinear polyatomic ions.

The equation for lattice potential energy U_{POT} has the form:

$$U_{POT}\ [\text{kJ mol}^{-1}] = x(\rho/M)^{1/3} + y \quad (3)$$

where $\rho/\text{g cm}^{-3}$ is the density, M is the chemical formula mass of the ionic material, and values for the coefficients $x/\text{kJ mol}^{-1}\ \text{cm}$ and $y/\text{kJ mol}^{-1}$ are taken from the literature [36].

4. Conclusions

In summary, through metathesis reactions between AgN_5 and corresponding chloride salts, six novel *cyclo*-N_5^- salts were successfully synthesized and comprehensively characterized. Their crystal structures were confirmed through single-crystal X-ray diffraction analyses. Physicochemical and energetic properties were also fully evaluated.

All *cyclo*-N_5^- salts exhibit superior safety characteristics with high mechanical stabilities (IS > 35 J, FS > 360 N), outperforming previous *cyclo*-N_5^- salts $N(CH_3)_4N_5$, $[DAG]N_5$, and $[EDA](N_5)_2$. Thermal analyses in this study demonstrated that **1–6** exhibit higher decomposition temperatures ($T_{peak} \geq 107$ °C) compared to $N(CH_3)_4N_5$, $[DAG]N_5$, and $[EDA](N_5)_2$. These enhanced stabilities can be attributed to their dense hydrogen-bonding networks. Furthermore, all *cyclo*-N_5^- salts demonstrate positive heats of formation in the range of 210.33–533.86 kJ mol^{-1}, detonation velocities of 7122–8288 m s^{-1}, and detonation pressures of 14.6–25.1 GPa. This work provides valuable insights for crystal engineers to further enhance the stability of *cyclo*-N_5^- salts and explore a broader variety of energetic materials containing *cyclo*-N_5^-.

Supplementary Materials: The following supporting information can be downloaded at: https://www.mdpi.com/article/10.3390/molecules30122613/s1, ^1H and ^{13}C NMR spectra; crystal structure data (PDF).

Author Contributions: Conceptualization, Y.X.; Methodology, Y.X.; Validation, G.L.; Formal analysis, H.Z.; Investigation, L.S. and S.J.; Resources, M.L. and Y.X.; Data curation, X.Y.; Writing—original draft, L.S.; Writing—review & editing, Y.X.; Supervision, Y.X.; Project administration, M.L.; Funding acquisition, M.L. and Y.X. All authors have read and agreed to the published version of the manuscript.

Funding: This work was supported by the National Natural Science Foundation of China (no. 22475102 and 22135003) and Young Elite Scientist Sponsorship Program by CAST (no. YESS20210074).

Institutional Review Board Statement: Not applicable.

Informed Consent Statement: Not applicable.

Data Availability Statement: The data presented in this study are available in the paper.

Conflicts of Interest: The authors declare no conflicts of interest.

References

1. Klapötke, T.M.; Piercey, D.G. 1,1′-azobis(tetrazole): A highly energetic nitrogen-rich compound with a N_{10} chain. *Inorg. Chem.* **2011**, *50*, 2732–2734. [CrossRef] [PubMed]
2. Joo, Y.H.; Twamley, B.; Garg, S.; Shreeve, J.M. Energetic nitrogen-rich derivatives of 1,5-diaminotetrazole. *Angew. Chem. Int. Ed.* **2008**, *47*, 6232–6239. [CrossRef] [PubMed]
3. Wang, Q.; Pang, F.; Wang, G.; Huang, J.; Nie, F.; Chen, F.X. Pentazadiene: A high-nitrogen linkage in energetic materials. *Chem. Commun.* **2017**, *53*, 2327–2330. [CrossRef] [PubMed]
4. Chavez, D.E.; Hiskey, M.A.; Gilardi, R.D. 3,3′-Azobis(6-amino-1,2,4,5-tetrazine): A novel high-nitrogen energetic material. *Angew. Chem. Int. Ed.* **2000**, *112*, 1861–1863. [CrossRef]
5. Chen, D.; Yang, H.; Yi, Z.; Xiong, H.; Zhang, L.; Zhu, S.; Cheng, G. $C_8N_{26}H_4$: An Environmentally Friendly Primary Explosive with High Heat of Formation. *Angew. Chem. Int. Ed.* **2018**, *130*, 2103–2106. [CrossRef]
6. Klapötke, T.M.; Sabaté, C.M. Mater. Nitrogen-rich tetrazolium azotetrazolate salts: A new family of insensitive energetic materials. *Chem. Mater.* **2008**, *20*, 1750–1763. [CrossRef]

7. Klapötke, T.M.; Martin, F.A.; Stierstorfer, J. An energetic and highly sensitive binary azidotetrazole. *Angew. Chem. Int. Ed.* **2011**, *50*, 4227. [CrossRef]
8. Shlomovich, A.; Pechersky, T.; Cohen, A.; Yan, Q.L.; Kosa, M.; Petrutik, N.; Tal, N.; Aizikovich, A.; Gozin, M. Energetic isomers of 1,2,4,5-tetrazine-bis-1,2,4-triazoles with low toxicity. *Dalton Trans.* **2017**, *46*, 5994–6002. [CrossRef]
9. Singh, R.P.; Verma, R.D.; Meshri, D.T.; Shreeve, J.M. Energetic nitrogen-rich salts and ionic liquids. *Angew. Chem. Int. Ed.* **2006**, *45*, 3584–3601. [CrossRef]
10. Gao, H.; Shreeve, J.M. Azole-Based Energetic Salts. *Chem. Rev.* **2011**, *111*, 7377–7436. [CrossRef]
11. Klapötke, T.M.; Witkowski, T.G. Nitrogen-Rich Energetic 1,2,5-Oxadiazole-Tetrazole–Based Energetic Materials. *Propell. Explos. Pyrotech.* **2015**, *40*, 366–373. [CrossRef]
12. Wozniak, D.R.; Piercey, D.G. Review of the current synthesis and properties of energetic pentazolate and derivatives thereof. *Engineering* **2020**, *6*, 981–991. [CrossRef]
13. Wang, P.; Xu, Y.; Lin, Q.; Lu, M. Recent advances in the syntheses and properties of polynitrogen pentazolate anion *cyclo*-N_5^- and its derivatives. *Chem. Soc. Rev.* **2018**, *47*, 7522–7538. [CrossRef] [PubMed]
14. Yang, C.; Zhang, C.; Zheng, Z.; Jiang, C.; Luo, J.; Du, Y.; Hu, B.; Sun, C.; Christe, K.O. Synthesis and characterization of *cyclo*-pentazolate salts of NH_4^+, NH_3OH^+, $N_2H_5^+$, $C(NH_2)_3^+$, and $N(CH_3)_4^+$. *J. Am. Chem. Soc.* **2018**, *140*, 16488–16494. [CrossRef]
15. Laniel, D.; Weck, G.; Gaiffe, G.; Garbarino, G.; Loubeyre, P. High-pressure synthesized lithium pentazolate compound metastable under ambient conditions. *J. Phys. Chem. Lett.* **2018**, *9*, 1600–1604. [CrossRef]
16. Yang, C.; Chen, L.; Wu, W.; Zhang, C.; Sun, C.; Du, Y.; Hu, B. Investigating the Stabilizing Forces of Pentazolate Salts. *ACS Appl. Energy Mater.* **2021**, *4*, 146–153. [CrossRef]
17. Li, X.; Long, Y.; Zhang, C.; Sun, C.; Hu, B.; Lu, P.; Chen, J. Symmetrical *cyclo*-N_5^- hydrogen bonds: Stabilization mechanism of four non-metallic *cyclo*-pentazolate energetic salts. *Phys. Chem. Chem. Phys.* **2022**, *24*, 3970–3983. [CrossRef]
18. Xu, Y.; Lin, Q.; Wang, P.; Lu, M. Syntheses, crystal structures and properties of a series of 3D metal-inorganic frameworks containing pentazolate anion. *Chem. Asian J.* **2018**, *13*, 1669–1673. [CrossRef]
19. Xu, Y.; Tian, L.; Li, D.; Wang, P.; Lu, M. A series of energetic *cyclo*-pentazolate salts: Rapid synthesis, characterization, and promising performance. *J. Mater. Chem. A* **2019**, *7*, 12468–12479. [CrossRef]
20. Xu, Y.; Wang, Q.; Shen, C.; Lin, Q.; Wang, P.; Lu, M. A series of energetic metal pentazolate hydrates. *Nature* **2017**, *549*, 78–81. [CrossRef]
21. Zhang, C.; Yang, C.; Hu, B.; Yu, C.; Zheng, Z.; Sun, C. Asymmetric $Co(N_5)_2(H_2O)_4{\cdot}4H_2O$ high-nitrogen compound formed by Cobalt (II) cation trapping of a *cyclo*-N_5^- anion. *Angew. Chem. Int. Ed.* **2017**, *56*, 4512–4514. [CrossRef] [PubMed]
22. Tian, L.; Xu, Y.; Lin, Q.; Wang, P.; Lu, M. Syntheses of energetic *cyclo*-pentazolate salts. *Chem. Asian J.* **2019**, *14*, 2877–2882. [CrossRef] [PubMed]
23. Xia, H.; Qi, X.; Zhang, W.; Huang, S.; Song, S.; Liu, Y.; Luo, J.; Zhang, Q. New Insight into the Aromaticity of *cyclo*-N_5^- by Constructing 3D Arrays in Crystal Structures. *Cryst. Growth Des.* **2021**, *21*, 33–39. [CrossRef]
24. Turner, M.J.; McKinnon, J.J.; Wolff, S.K.; Grimwood, D.J.; Spackman, P.R.; Jayatilaka, D.; Spackman, M.A. *CrystalExplorer17*; University of Western Australia: Perth, Australia, 2017.
25. Spackman, M.A.; Jayatilaka, D. Hirshfeld surface analysis. *CrystEngComm* **2009**, *11*, 19–32. [CrossRef]
26. Gao, H.; Ye, C.; Piekarski, C.M.; Shreeve, J.M. Computational Characterization of Energetic Salts. *J. Phys. Chem. C* **2007**, *111*, 10718–10731. [CrossRef]
27. Zhang, Y.; Huang, Y.; Parrish, D.A.; Shreeve, J.M. 4-Amino-3,5-dinitropyrazolate salts-highly insensitive energetic materials. *J. Mater. Chem. A* **2011**, *21*, 6891–6897. [CrossRef]
28. Yu, R.J.; Liu, Y.J.; Huang, W.; Tang, Y.X. A hybrid of tetrazolium and pentazolate: An energetic salt with ultrahigh nitrogen content and energy. *Energ. Mater. Front.* **2023**, *4*, 63–67. [CrossRef]
29. Chand, D.; Parrish, D.A.; Shreeve, J.M. Di(1*H*-tetrazol-5-yl)methanone oxime and 5,5′-(hydrazonomethylene)bis(1*H*-tetrazole) and their salts: A family of highly useful new tetrazoles and energetic materials. *J. Mater. Chem. A* **2013**, *1*, 15383–15389. [CrossRef]
30. Frisch, M.J.; Trucks, G.W.; Schlegel, H.B.; Scuseria, G.E.; Robb, M.A.; Cheeseman, J.R.; Scalmani, G.; Barone, V.; Mennucci, B.; Petersson, G.A.; et al. *Gaussian 09, Revision A.02*; Gaussian, Inc.: Wallingford, CT, USA, 2009.
31. Sućeska, M. *EXPLO5 V6.05.04*; Brodarski Institute: Zagreb, Croatia, 2020.
32. United Nations; Committee of Experts on the Transport of Dangerous Goods. *Recommendations on the Transport of Dangerous Goods: Manual of Tests and Criteria*; United Nations Publications: New York, NY, USA, 2009.
33. Stephens, P.J.; Devlin, F.J.; Chabalowski, C.F.; Frisch, M.J. Ab Initio Calculation of Vibrational Absorption and Circular Dichroism Spectra Using Density Functional Force Fields. *J. Phys. Chem.* **1994**, *98*, 11623–11627. [CrossRef]
34. Ochterski, J.W.; Petersson, G.A.; Montgomery, J.A. A complete basis set model chemistry. V. Extensions to six or more heavy atoms. *J. Chem. Phys.* **1996**, *104*, 2598–2619. [CrossRef]

35. Jenkins, H.D.B.; Tudela, D.; Glasser, L. Lattice potential energy estimation for complex ionic salts from density measurements. *Inorg. Chem.* **2002**, *41*, 2364–2367. [CrossRef]
36. Jenkins, H.D.B.; Roobottom, H.K.; Passmore, J.; Glasser, L. Relationships among ionic lattice energies, molecular (Formula Unit) volumes, and thermochemical radii. *Inorg. Chem.* **1999**, *38*, 3609–3620. [CrossRef] [PubMed]

molecules

MDPI

Article

1,1′-(Diazene-1,2-diyl)bis(4-nitro-1H-1,2,3-triazole-5-carboxamide): An N8-Type Energetic Compound with Enhanced Molecular Stability

Moxin Sun [1], Wenjie Xie [1], Qi Lai [1], Gang Zhao [2,3,*], Ping Yin [1,*] and Siping Pang [1]

[1] School of Materials Science and Engineering, Beijing Institute of Technology, Beijing 100081, China; 3120221080@bit.edu.cn (M.S.); 3220211405@bit.edu.cn (W.X.); laiqi@bit.edu.cn (Q.L.); pangsp@bit.edu.cn (S.P.)

[2] Sichuan Huachuan Industrial Co., Ltd., Chengdu 610106, China

[3] School of Materials Science and Engineering, Harbin Institute of Technology (Shenzhen), Shenzhen 518055, China

* Correspondence: zhaogang@hit.edu.cn (G.Z.); pingyin@bit.edu.cn (P.Y.)

Abstract: The safety concerns associated with sensitivity issues regarding long nitrogen chain-based energetic compounds, especially for eight or more catenated nitrogen atoms in backbones, need to be resolved. Incorporating specific functional groups represents a key approach for enhancing stability in organic energetic materials. This study reports the synthesis of 1,1′-(diazene-1,2-diyl)bis(4-nitro-1H-1,2,3-triazole-5-carboxamide) (**S8**), an N8-chain compound featuring strategically placed amide groups. Employing THA(*O*-tosylhydroxylamine) and $KMnO_4$, 1,1′-(diazene-1,2-diyl)bis(4-nitro-1H-1,2,3-triazole-5-carboxamide) (**S8**) was synthesized and underwent N-amination and oxidative azo coupling. Comprehensive characterization, including X-ray diffraction, mechanical sensitivity testing, and theoretical analysis, alongside comparative studies with known N8 compounds, revealed that S8 exhibits unprecedented stability within its class. Among reported N8-catenated nitrogen chain compounds, attributed to the incorporation of the amide functionality, **S8** demonstrates the highest impact sensitivity (IS = 10 J) and friction sensitivity (FS = 40 N) while maintaining excellent detonation performance (D = 8317 ms^{-1}, P = 28.27 GPa). This work highlights the amide group as a critical structural part for achieving high stability in sensitive long-nitrogen-chain energetic materials without compromising performance.

Keywords: long nitrogen chain compounds; nitrogen-rich heterocyclic ring; azo bridge; nitro-1,2,3-triazolecarboxamide; low-sensitivity energetic materials

1. Introduction

Energetic materials are regarded as a crucial part of advanced equipment which directly determines the development of civil applications and national defense. They are substances that can undergo intense oxidation–reduction reactions and release a large amount of energy under certain external stimuli. With the rapid development of energetic materials, higher requirements have been put forward for new energetic materials, including higher energy, greater security, and more environmental friendliness [1,2]. To meet these requirements, nitrogen-rich heterocycles like pyrazole, imidazole, triazole, tetrazole, and their derivatives have occupied a significant place in energetic materials in recent years [3], since they have a host of N–N and C–N bonds, exhibit high heats of formation, remarkable insensitivity, enhanced density, and environmental friendliness. Among them,

https://doi.org/10.3390/molecules30122589

long nitrogen chain heterocycles have aroused more interest due to their high heat of formation, which arises from the large energy release during the formation of N–N bonds [4]. However, materials with sufficient energy are often too sensitive to external mechanical stimuli. In contrast, many energetic materials with adequate stability do not meet the performance requirements [5]. Literature shows that long nitrogen chain compounds with both high energetic performance and good mechanical stability are rarely observed, and compounds become increasingly sensitive to external stimuli accompanied by the growth of the nitrogen–atom chain. According to the literature, the impact sensitivity of N6 is within the range of 2–14 J [6–8], the value of N8 is between 0.25 and 4 J [3,9–11], and the value of N10 is between 0 and 1 J [12].

Molecular design strategies that enhance electron delocalization and conjugation effects are critical for modifying material properties, as demonstrated in diverse chemical systems [13–15]. Notably, prior studies have not explored the introduction of functional groups incorporating both hydrogen-bond donors and acceptors (e.g., amides) into long-catenated nitrogen chain compounds to address their inherent instability. To address this gap, the amide group was strategically introduced into the N8-type long-nitrogen-chain molecule **S8**. Our experimental findings demonstrate that this strategic modification significantly enhances the stability of **S8**, establishing a promising approach to address sensitivity issues in long-nitrogen-chain energetic materials.

In this work, we report a simple method of HB introduction in triazole-based long nitrogen chain compound N8L to improve molecular stability. The compounds were characterized by infrared and multinuclear NMR spectroscopy. Additionally, single-crystal X-ray diffraction was performed to study the structural characteristics of intermolecular interactions (Figure 1).

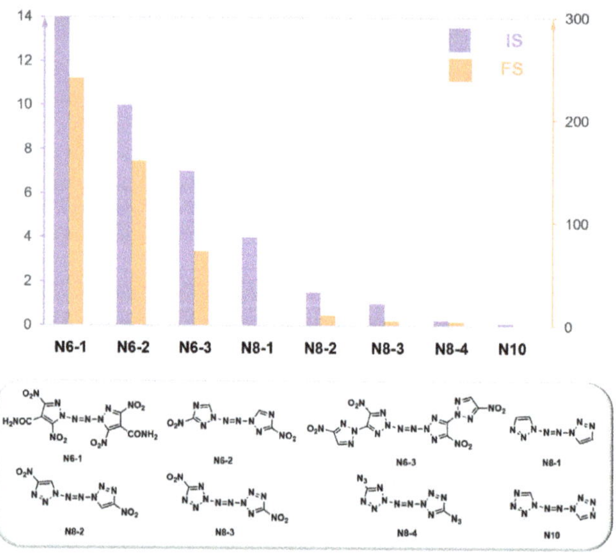

Figure 1. Mechanical sensitivity of long nitrogen chain compounds. N6-1 [8], N6-2 [7], N6-3 [6], N8-1 [4], N8-2 [9], N8-3 [10], N8-4 [11], N10 [12].

2. Results and Discussion

2.1. Synthesis

The synthetic procedures for long-chain N-rich compounds are given in Scheme 1. Compounds **1** and **2** were prepared according to the literature [16]. Further reactions involving the neutralization of compound **2** with DBU and the amination of compound **2** using freshly prepared o-p-toluenesulfonyl-hydroxylamine (THA) gave rise to compounds **3** in 51% yields. It is worth noting that the reported reactions of 1,2,3-triazole have rarely shown regioselectivity [17]. For example, the amination of diammonium 4,4′-bis(5-nitro 1,2,3-2H-triazolate) led to the formation of three isomers [17], and the amination of 4-nitro-2H-1,2,3-triazole led to the formation of two isomers [9]. However, no isomers of compound **3** could be observed in this reaction, and the selective N-amination could be rationalized from the steric effect of the amide group.

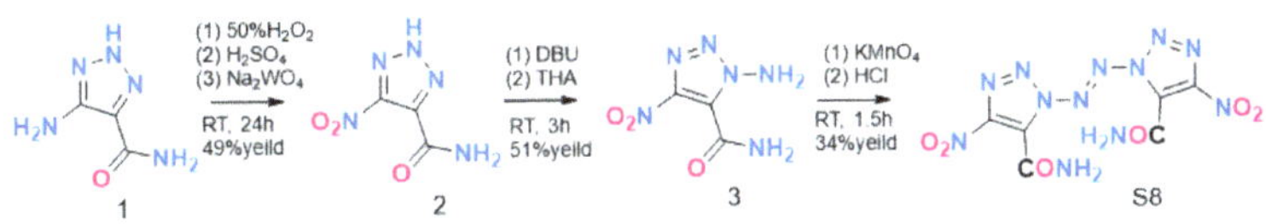

Scheme 1. Synthetic route of compound **S8**.

2.2. *Single Crystal X-Ray Structure Analysis*

To confirm the structures of **S8** and explore their HB interactions, crystals suitable for single-crystal X-ray diffraction were obtained from the slow evaporation of chloroform solution. Compound **S8** crystallizes in the monoclinic space group $P2_12_12_1$ with a calculated crystal density of 1.763 g cm^{-3} at 100 K (Z = 4). The crystal packing of S8 can be viewed as a face-to-face wavelike stacking with a 3.608 Å interlayer spacing, within the range of π–π interaction (<4.0 Å) [13] (Figure 2b). The packing index is 70.44%, which is a desirable packing type that contributes to low sensitivity. The bond lengths of C6-NO2 and C1-NO2 are 1.448(6) and 1.427(6) Å, respectively. There are four molecules in the **S8** unit cell, and each molecule has a near-planar geometry, which can be further confirmed by the torsion angles of [N7-N8-N9-C6 = −0.1 (5)°, N5-N4-N3-N2 = 179.7 (4)°, C2-N4-N3-N2 = 0.3(6)°, N3-N2-C1-N1 = 179.8(4)°], and the torsion angles of the nitro group (C6-NO2)[O5-N10-C6-C4 = 8.6(8)°], (C1-NO2)[C2-C1-N1-O1 = −11.7(7)°].

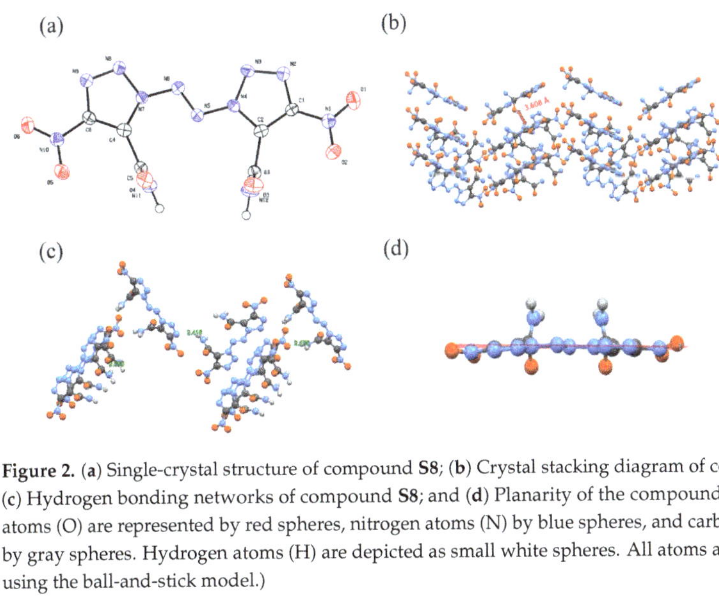

Figure 2. (**a**) Single-crystal structure of compound **S8**; (**b**) Crystal stacking diagram of compound **S8**; (**c**) Hydrogen bonding networks of compound **S8**; and (**d**) Planarity of the compound **S8**. (oxygen atoms (O) are represented by red spheres, nitrogen atoms (N) by blue spheres, and carbon atoms (C) by gray spheres. Hydrogen atoms (H) are depicted as small white spheres. All atoms are illustrated using the ball-and-stick model.)

X-ray single-crystal analysis revealed that each **S8** molecule participates in eight strong intermolecular hydrogen bonds toward the nitrogen of the triazole ring, oxygen and hydrogen of the amide, and the oxygen of the nitro group (Figure 3). Red spots on the **S8** Hirshfeld surface (Figure 3a) are large and colorful, indicating that the intermolecular hydrogen bonds in **S8** are plentiful and strong. Each molecule within the crystal structure participates in 8 intermolecular hydrogen bonds. The above analysis shows 3D HB networks on **S8**.

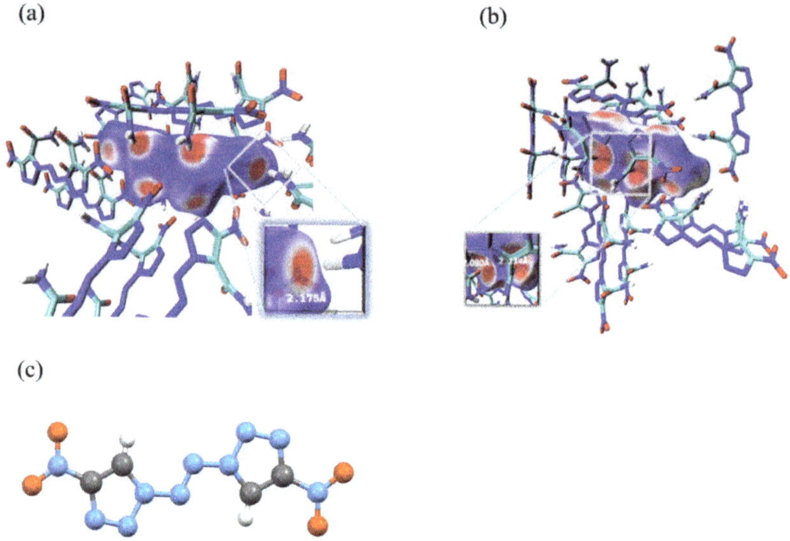

Figure 3. Hirshfeld surfaces for compound **S8** (**a**,**b**); Given the very similar structure. In subsequent chapters, frequent comparative analysis was conducted on **S8** and **N8L** to observe whether their performance was improved. The structure of **N8L** is shown in (c). (oxygen atoms (O) are represented by red spheres, nitrogen atoms (N) by blue spheres, and carbon atoms (C) by gray spheres. Hydrogen atoms (H) are depicted as small white spheres. All atoms are illustrated using the ball-and-stick model).

Two-dimensional fingerprints of **S8** and the associated Hirshfeld surface were analyzed. Red spots in **S8** are more extensive than those of **N8L**, indicating that the intermolecular hydrogen bonds in **S8** are stronger and more plentiful than those of **N8L**. Red spots in **S8** are on the same side of the molecular, which indicates that the intermolecular interaction mainly occurs on the side with the functionalized group. **S8** possesses strong hydrogen-bonding interactions, which can be rationalized from the strong O. . .H spikes in the Hirshfeld surface (Figure 4c). Quantifying this in Figure 4a,b, 43.5% of interactions in **S8** are N. . .H and O. . .H bonds, while for **N8L,** this value is only 24%.

Electrostatic potential surfaces (ESPs) were calculated to compare the difference in physicochemical properties [19–21]. The maximum value of **N8L** (+53.72 kcal mol^{-1}) is higher than that of **S8** (+51.96 kcal mol^{-1}). It is worth noting that ESP values may be related to mechanical sensitivities [22,23]. The number of **S8** is 80.48 kcal mol^{-1}, which is larger than that of **N8L** (79.91 kcal mol^{-1}) (Figure 5).

To investigate the influence of the introduction of amide droup, the density, detonation performance, thermal stability, mechanical sensitivity, and heat of formation of **S8** were studied and compared with other N8 compounds. We tested the mechanical sensitivity of **S8** using the BFH 10 model BAM drop hammer impact sensitivity apparatus and FKSM 10BAM model BAM friction sensitivity apparatus at room temperature. The impact sensitivities (*IS*) and friction sensitivities (*FS*) of **S8** were *IS* = 10 J, and *FS* = 40 N, which is much less sensitive than that of other N8-type energetic compounds. According to the study of crystal structures and comparison with **N8L**, we propose that the introduction of amide is responsible for the improvement in the mechanical sensitivity towards external mechanic stimuli. Other reported long-catenated nitrogen structures are listed below, and detailed sensitivity and detonation performance are drawn in Figure 6.

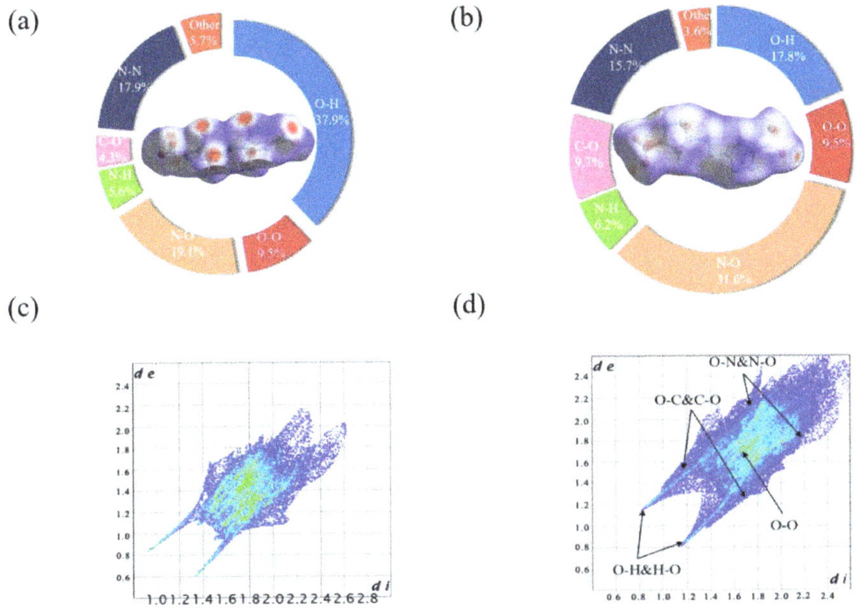

Figure 4. Quantitative analysis of intermolecular hydrogen bonds of **S8** and **N8L**. Hirshfeld [18] surfaces (inside) and p ie graphs of **S8** (**a**) and **N8L** (**b**), respectively, showing the percentage contribution of the individual atomic contacts to the Hirshfeld surfaces (**c**,**d**). Highlighting N. . .H and O. . .H contacts (inside) and 2D-fingerprint plots in crystal stacking, respectively.

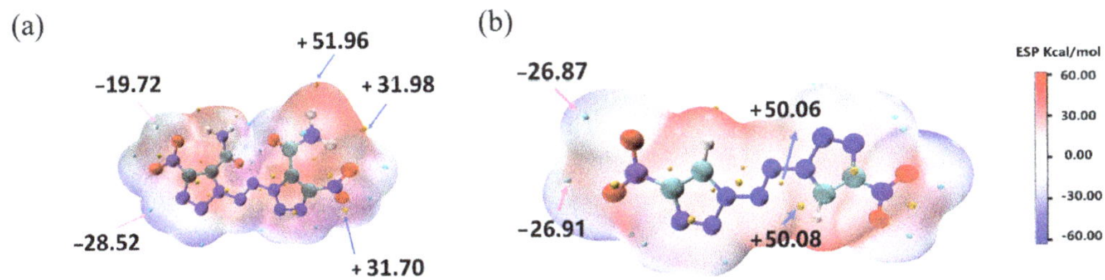

Figure 5. Electrostatic potential surfaces (ESPs) for **S8** (**a**) and **N8L** (**b**), respectively. The blue and red spheres in the diagram represent the surface local minima and maximum of ESP, respectively.

Based on isodesmic reactions and calculated by the Gaussian 09 D.01 program, **S8** has a fine heat of formation of 687.45 kJ mol^{-1}, which can be attributed to the large number of N=N and N-N bonds in the molecule. Measuring by Gas pycnometers, the density of **S8** in 100k was 1.767 g cm^{-3}. Key parameters were selected to calculate with EXPLO5 (version 6.05) to estimate the detonation performances. As a result, **S8** has a high detonation velocity and pressure with the calculated number of 8317 ms^{-1} and 33.43 GPa, respectively.

Differential scanning calorimetry and thermal gravimetric results reveal that **S8** possesses higher thermostability, with a T_d of 177.6 °C. It is worth noting that the introduction of amide causes an improvement of 1 °C in thermal stability. It is also worth noting that this system represents the first example of the introduction of the amide group to improve molecular stability in the field of long nitrogen chain compounds (Table 1).

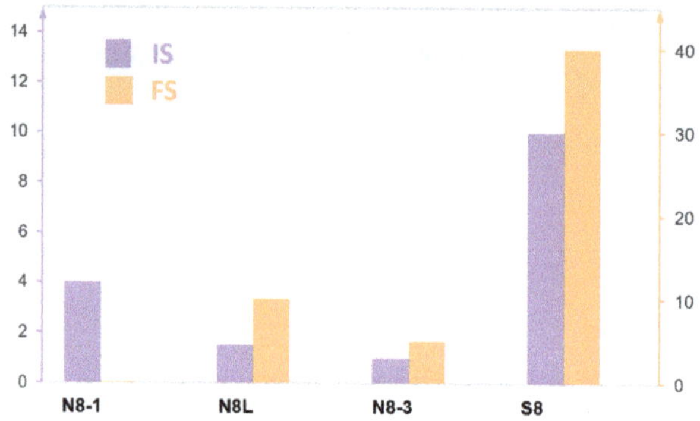

Figure 6. Comparison of sensitivity (Friction sensitivity and impact sensitivity) between **S8** and compounds **N8-1** [4], **N8L** [9], **N8-3** [10].

Table 1. The physical properties of compounds **S8** compared with other N8 compounds.

Comp.	T_d [a] [°C] [a]	d [b] [gcm^{-3}] [b]	$\Delta_f H$ [c] [KJ mol^{-1}] [c]	IS [d] [J] [d]	FS [e] [N] [e]	D [f] [ms^{-1}] [f]	P [g] [GPa]
N8L	177	1.78	877.7	1.5	10	8916	33.5
S8	177.6	1.76	687.45	10	40	8317	28.27
N8-1 [h]	193.8	1.62	962	4	-	7764	25.24
N8-3 [i]	-	1.80	1153	<<<<1	<<<<5	9184	39
N8-4 [j]	114	1.81	1700	<<0.25	<<0.1	9515	36.6
RDX	204	1.80	70.3	7.4	120	8762	18.4
TNT	295	1.65	69.75	2.5–4	0.1–1	7190	34.9

[a] Thermal decomposition temperature, °C. [b] Crystal density at 296 K, g cm^{-3}. [c] Heats of formation (computed by the Gaussian 09 D.01 program [24]), kJ mol^{-1}. [d] Impact sensitivity, J. [e] Friction sensitivity, N. [f] Detonation velocity (calculated with Explo5 v6.02). [g] Detonation pressure, GPa. [h] Ref. [4]. [i] Ref. [10]. [j] Ref. [11].

3. Materials and Methods

Caution! The novel compounds investigated in this work possess inherent energetic properties, creating a potential for detonation under specific external stimuli. Consequently, all experimental procedures involving these materials demand enhanced safety protocols. Prudent measures include the consistent use of protective shields, hearing protection (earplugs), safety goggles, and cut-resistant gloves.

3.1. Reagents and Instruments

Reagents: 3-Amino-3-iminopropanamide hydrochloride (1:1), sodium methoxide, p-toluenesulfonyl azide, sodium tungstate dihydrate, 50% hydrogen peroxide solution, dichloromethane, anhydrous sodium sulfate, ethyl acetate, triethylamine, ethyl aceto-hydroxamic acid, p-toluenesulfonyl chloride, perchloric acid, potassium permanganate, concentrated hydrochloric acid and acetonitrile all of analytical grade, were purchased from Shanghai Aladdin Biochemical Technology Co., Ltd, Beijing, China. Energy-chemical, Shanghai, China. Beijing Chemical Plant Co., Ltd. Beijing, China, Beijing Tong Guang Fine Chemcials Company, Beijing, China. Shanghai Meiruier Biochemical Technology Co., Ltd, Shanghai, China.

Instruments: AVANCE DRX-500 NMR scanner, Bruker, Fällanden, Switzerland; RCT basic magnetic stirrer, IKA Staufen im Breisgau, Germany; NI 10 Infrared Spectrometer, Thermo Fisher Scientific, Waltham, MA, USA; Q2000 Differential Thermal Scanner, US TA, New Castle, DE, USA; FLASH 2000 CHNS/O Element Analyzer, Thermo Fisher Scientific; BFH 10 BAM Drop Hammer Impact Sensitivity Instrument, FKSM 10 BAM Friction Sensitivity Tester, Edison Company, Beijing, China.

3.2. Computational Methods

- Electrostatic potential (ESP) and Heat of Formation

The geometric optimization of compounds **3** and **S8** was carried out using Gaussian 09, density functional theory (DFT), and the B3LYP/6-31g (d, p) basis set. The vibration analysis of the optimized structures showed no imaginary frequencies, indicating that these structures are at their minimum values on their respective potential energy surfaces. Subsequently, the electrostatic potentials of compounds **3** and **S8**, as well as the enthalpy of formation of compound **S8**, were calculated and analyzed using the B3LYP/6-31+g (d, p) basis set.

3.3. Compound 1,2 Prepared According to the Literature [25]

- Synthesis of compound **1**:

Malonamamidine hydrochloride (0.1 mol, 41.28 g) was added to a sodium methoxide solution (3 N, 100 mL) at 0 °C for reaction. After the neutralization reaction, the precipitate was isolated by filtering the final mixture and washed with ethanol (40 mL). To this filtrate, an ethanol solution of p-methylbenzenesulfonyl azide was added dropwise, and the mixture was stirred overnight at ambient temperature. After the reaction, the mixture was filtered, and the precipitate was washed with ethanol and then air-dried. Yield: white solid (80%). ^{1}H NMR (d_6-DMSO): δ7.40 (s, 1 H), 7.08 (s, 1 H), 5.79 (s, 2 H) ppm. ^{13}C NMR (d_6-DMSO): δ 157.5, 147.3, 137.5, 129.4 ppm. IR (KBr): ṽ:3392, 1690, 1672, 1535, 1340, 1130, 1002, 850, 568 cm^{-1}. Elemental analysis of $C_3H_5N_5O$ (127.05): Calculated (%) C 28.35, H 3.97, N 55, O 12.59; found (%): C 28.01, H 4.10, N 54.5, O 13.39.

- Synthesis of compound **2**:

At 0 °C, Sodium tungstate dihydrate (6.5 g, 22.5 mmol) was dissolved in 50% H_2O_2 (100 mL), and concentrated sulfuric acid (0.5 mL) was added dropwise (the solution color is yellow at this time). Maintain the mixture at 0–10 °C, and the above 5-amino-1,2,3-triazol-4-formamide (3.14 g, 24.7 mmol) was added. After stirring for s days at ambient temperature, the mixture was filtered, and the precipitate was washed with H_2O and dried in air. Yield: yellow solid (49%). ^{1}H NMR (d^6-DMSO): δ8.27 (s, 1 H), 8.16 (s, 1 H) ppm. ^{13}C NMR (d_6-DMSO): δ 163.5, 151.7, 111.4, 108.5 ppm. IR (KBr): ṽ: 3452, 2980, 1686, 1597, 1498, 1416, 1402, 1355, 1305, 1277, 1183, 953, 849 cm^{-1}. Elemental analysis of $C_3H_3N_5O_3$ (157.02): Calculated (%) C 22.94, H 1.93, N 44.58, O 30.55; found (%): C 23.12, H 1.95, N 44.26, O 30.67.

- Synthesis of compound **3**:

Compound 2 (1570 mg, 10 mmol) was dissolved in 1,8-Diazabicyclo [5.4.0] undec-7-ene (1.57 mL, 10 mmol) at ambient temperature. After stirring for 30 min, 8 mL of freshly prepared O-Tosylhydroxylamine was added dropwise to this solution. After stirring for 3 h at ambient temperature, the mixture was filtered, and the precipitate was washed with H_2O and dried in air. Purification was performed by column chromatography using a petroleum ether/ethyl acetate mixture (1:2, v/v) as the eluent. (40%). ^{1}H NMR (d_6-DMSO) δ 8.360 (s, 2 H), 7.509 (s, 2 H). ^{13}C NMR (d_6-DMSO) δ 156.97, 147.01, 129.95 ppm. IR (KBr): ṽ:3272, 1694, 1526, 1387, 1262, 1201, 1031, 1011, 846, 793, 715, 683, 666, 528, 472 cm^{-1}. Elemental analysis of $C_3H_4N_6O_3$ (172.03): Calculated (%) C 20.94, H 2.34, N 48.83, O 27.89; found (%): C 21.12, H 2.22, N 48.69, O 27.97.

- Synthesis of compound **S8**:

Compound 3 (344 mg, 2 mmol) was added to a round-bottomed flask with 5 ml hydrochloric acid and 16 mL of $KMnO_4$ (500 mg, 3 mmol) solution was added dropwise to this solution at ambient temperature. After stirring for 1.5 h at ambient temperature, the mixture was filtered, and the precipitate was washed with H_2O and hydrochloric acid, then dried in the air. Yield: white solid (34%) ^{1}H NMR (d_3-CD_3CN.): δ 7.133 (br, 2H), δ 7.045 (s, 2H). ^{13}C NMR (d_3-CD_3CN): δ 181.296, 155.307, 130.311 ppm. IR (KBr): ṽ:1703, 1526, 1305, 908, 844, 831, 555, 533, 504 cm^{-1}. Elemental analysis of $C_6H_4N_{12}O_6$ (340.04): Calculated (%) C 21.18, H 1.19, N 49.41, O 28.22; found (%): C 20.96, H 1.38, N 49.01, O 28.65.

3.4. Single-Crystal Preparation Method of 3 and S8

Single crystals of novel compounds **3** and **S8** were grown using the solvent evaporation method. A single crystal measuring 0.19 mm × 0.18 mm × 0.16 mm was selected to ensure the acquisition of high-quality diffraction data during the X-ray diffraction analysis. Data collection: APEX-III; cell refinement: SAINT V8.40A (Bruker, 2019); data reduction: SAINT V8.40A (Bruker, 2019); program(s) used to solve structure: SHELXT 2014/5 (Sheldrick, 2014, Nairobi, Kenya); program(s) used to refine structure: SHELXL2019/2 (Sheldrick, 2019). Supplementary Materials provide detailed information on crystallography. Subsequent analysis of the spatial configurations was performed using visualization software to elucidate the structure–property relationships.

3.5. Thermal Performance and FTIR Experimental Methods

- Differential scanning calorimetry experiments

For DSC measurements, samples (~0.5 mg) were accurately weighed using an analytical balance, loaded into hermetically sealed aluminum pans, and analyzed under a dry nitrogen purge flow of 20 mL·min^{-1}. The temperature was ramped from 40 °C to 400 °C at a heating rate of 10 °C·min^{-1}.

- In situ FTIR experiments

Fourier-transform infrared (FT-IR) spectra were obtained at 25 °C using a PerkinElmer Spectrum BX spectrometer equipped with an attenuated total reflection (ATR) accessory.

4. Conclusions

The stabilization of high-energy materials enables them to function across multiple technological domains. Through enhanced mechanical sensitivity, these advanced formulations allow for reliable deployment in extreme environments such as Arctic/desert combat zones. Most importantly, such research contributes to the development of the next generation of high-energy insensitive explosives. This study successfully synthesized **S8**, an N8-type long-nitrogen-chain compound. Experimental results and data analysis demonstrate that the incorporation of an amide group enhances the stability of long-nitrogen-chain molecules. **S8** exhibits the highest stability among all reported N8-type long-nitrogen-chain compounds to date. This work provides deeper insights into the relationship between molecular structure and performance. Furthermore, it offers valuable perspectives for designing energetic materials with superior properties, particularly low sensitivity.

Supplementary Materials: The following supporting information can be downloaded at: https://www.mdpi.com/article/10.3390/molecules30122589/s1, Figure S1: Isodesmic reactions for **S8**; Table S1: Crystal data for compounds **3**; Table S2: Refinement for compounds **3**; Table S3: Data collection for compounds **3**; Table S4: Fractional atomic coordinates and isotropic or equivalent isotropic displacement parameters (Å^2) for compound **3**; Table S5: Atomic displacement parameters (Å^2) for compound **3**; Table S6: Geometric parameters (Å^2) for compound **3**; Table S7: Crystal data for compounds **S8**; Table S8: Refinement for compounds **S8**; Table S9: Data collection for compounds **S8**; Table S10: Fractional atomic coordinates and isotropic or equivalent isotropic displacement parameters (Å^2) for compound **S8**; Table S11: Atomic displacement parameters (Å^2) for **S8**; Table S12: Geometric parameters (Å^2) for **S8**; Figure S2: ^{1}H NMR spectrum of compound **3** in DMSO-*d6*; Figure S3: ^{13}C NMR spectrum of compound **3** in DMSO-*d6*; Figure S4: ^{1}H NMR spectrum of **S8** in *d3*-CD_3CN; Figure S5: ^{13}C NMR spectrum of **S8** in *d3*-CD_3CN; Figure S6: IR spectrum of compound **3**; Figure S7: IR spectrum of compound **S8**; Figure S8: DSC curve of compound **S8** at 10 °C min^{-1} [26–28].

Author Contributions: Conceptualization, G.Z. and S.P.; Methodology, M.S., W.X. and P.Y.; Writing—review & editing, Q.L. All authors have read and agreed to the published version of the manuscript.

Funding: This research received no external funding.

Institutional Review Board Statement: Not applicable.

Informed Consent Statement: Not applicable.

Data Availability Statement: CCDC 2362107 (**S8**), 2362106 (**3**) contain the supplementary crystallographic data for this paper. These data can be obtained free of charge via www.ccdc.cam.ac.uk/data_request/cif (accessed on 1 June 2025), or by emailing data_request@ccdc.cam.ac.uk, or by contacting The Cambridge Crystallographic Data Centre, 12 Union Road, Cambridge CB2 1EZ, UK; fax: +44 1223 336033.

Conflicts of Interest: Author Gang Zhao was employed by the company Sichuan Huachuan Industrial Co., Ltd. The remaining authors declare that the research was conducted in the absence of any commercial or financial relationships that could be construed as a potential conflict of interest.

References

1. He, C.; Imler, G.H.; Parrish, D.A.; Shreeve, J.M. Energetic salts of 4-Nitramino-3- (5-dinitromethyl-1,2,4-oxadiazolyl)-furazan: Powerful alliance towards good thermal stability and high performance. *J. Mater. Chem. A* **2018**, *6*, 16833–16837. [CrossRef]
2. Ma, J.; Chinnam, A.K.; Cheng, G.; Yang, H.; Zhang, J.; Shreeve, J.M. 1,3,4- oxadiazole bridges: A strategy to improve energetics at the molecular level. *Angew. Chem. Int. Ed.* **2021**, *60*, 5497–5504. [CrossRef]
3. Dalinger, I.L.; Kormanov, A.V.; Suponitsky, K.Y.; Muravyev, N.V.; Sheremetev, A.B. Pyrazole–tetrazole hybrid with trinitromethyl, fluorodinitromethyl, or (difluoroamino) dinitromethyl groups: High-performance energetic materials. *Chem. Asian J.* **2018**, *13*, 1165–1172. [CrossRef] [PubMed]
4. Li, Y.; Qi, C.; Li, S.; Zhang, H.; Sun, C.; Yu, Y.; Pang, S. 1,1′-Azobis-1,2,3-triazole: A high-nitrogen compound with stable N8 structure and photochromism. *J. Am. Chem. Soc.* **2010**, *132*, 12172–12173. [CrossRef]
5. Dippold, A.A.; Klapötke, T.M. A study of dinitro-bis-1,2,4-triazole-1,1′-diol and derivatives: Design of high-performance insensitive energetic materials by the introduction of N-oxides. *J. Am. Chem. Soc.* **2013**, *135*, 9931–9938. [CrossRef]
6. Feng, S.; Li, Y.; Lai, Q.; Wang, Z.; Yin, P.; He, C.; Pang, S. A strategy for stabilizing of n8 type energetic materials by introducing 4-nitro-1,2,3-triazole scaffolds. *Chem. Eng. J.* **2022**, *430*, 133181. [CrossRef]
7. Yin, P.; Parrish, D.A.; Shreeve, J.M. N-diazo-bridged nitroazoles: Catenated nitrogen-atom chains compatible with nitro functionalities. *Chem. Eur. J.* **2014**, *20*, 6707–6712. [CrossRef]
8. Xie, W.; Liu, Y.; Zhou, M.; Yin, P.; Pang, S. 1,2,3-triazole with linear and branched catenated nitrogen chains—The role of regiochemistry in energetic materials. *Chem. Eng. J.* **2024**, *488*, 150974. [CrossRef]
9. Lai, Q.; Fei, T.; Yin, P.; Shreeve, J.M. 1,2,3-triazole with linear and branched catenated nitrogen chains—The role of regiochemistry in energetic materials. *Chem. Eng. J.* **2021**, *410*, 128148. [CrossRef]
10. Klapötke, T.M.; Piercey, D.G.; Stierstorfer, J. Amination of energetic anions: High-performing energetic materials. *Dalton Trans.* **2012**, *41*, 9451–9459. [CrossRef]
11. Benz, M.; Klapötke, T.M.; Stierstorfer, J.; Voggenreiter, M. Synthesis and characterization of binary, highly endothermic, and extremely sensitive 2,2′-azobis(5-azidotetrazole). *J. Am. Chem. Soc.* **2022**, *144*, 6143–6147. [CrossRef] [PubMed]
12. Klapötke, T.M.; Piercey, D.G. ; 1,10-Azobis(tetrazole): A Highly Energetic Nitrogen-Rich Compound with a N10 Chain. *Inorg. Chem.* **2011**, *50*, 2732–2734. [CrossRef] [PubMed]

13. Sun, Q.; Li, X.; Bamforth, C.; Lin, Q.; Murugesu, M.; Lu, M. Higher performing and less sensitive CN_7^- -based high-energy-density material. *Sci. Chi. Mater.* **2020**, *63*, 1779–1787. [CrossRef]
14. Cai, Y.; Hai, Y.; Ohashi, M.; Jamieson, C.S.; Garcia-Borras, M.; Houk, K.N.; Zhou, J.; Tang, Y. Structural basis for stereoselective dehydration and hydrogen-bonding catalysis by the SAM-dependent pericyclase LepI. *Nature Chem.* **2019**, *11*, 812–820. [CrossRef]
15. James, N.M.; Han, E.; Cruz, R.A.L.; Jureller, J.; Jaeger, H.M. Interparticle hydrogen bonding can elicit shear jamming in dense suspensions. *Nature Mater.* **2018**, *17*, 965–970. [CrossRef] [PubMed]
16. Schawartz, J.; Hornyák, M.; Süts, T. Preparation of 5-amino-1H,1,2,3-triazole-4-carboxamide. *Chem. Ind. London.* **1970**, *3*, 93.
17. He, C.; Shreeve, J.M. Energetic Materials with Promising Properties: Synthesis and characterization of 4,4′-bis(5-nitro-1,2,3-2h-triazole) derivatives. *Angew. Chem. Int. Ed.* **2015**, *54*, 6260–6264. [CrossRef]
18. Spackman, P.R.; Turner, M.J.; McKinnon, J.J.; Wolff, S.K.; Grimwood, D.J.; Jayatilaka, D.; Spackman, M.A. CrystalExplorer: A program for Hirshfeld surface analysis, visualization and quantitative analysis of molecular crystals. *J. Appl. Crystallogr.* **2021**, *54*, 1006–1011. [CrossRef]
19. Lu, T.; Chen, F. Multiwfn: A multifunctional wavefunction analyzer. *J. Comput. Chem.* **2012**, *33*, 580–592. [CrossRef]
20. Humphrey, W.; Dalke, A.; Schulten, K. VMD: Visual molecular dynamics. *J. Mol. Graph. Model.* **1996**, *14*, 33–38. [CrossRef]
21. Lu, T.; Manzetti, S. Wavefunction and reactivity study of benzo[a]pyrene diol epoxide and its enantiomeric forms. *Struct. Chem.* **2014**, *25*, 1521–1533. [CrossRef]
22. Geng, W.; Ma, Q.; Chen, Y.; Yang, W.; Jia, Y.; Li, J.; Zhang, Z.; Fan, G.; Wang, S. Structureperformance relationship in thermally stable energetic materials: Tunable physical properties of benzopyridotetraazapentalene by incorporating amino groups, hydrogen bonding, and π-π interactions. *Cryst. Growth Des.* **2020**, *20*, 2106–2114. [CrossRef]
23. Zhang, W.; Zhang, J.; Deng, M.; Qi, X.; Nie, F.; Zhang, Q. A promising high-energy density material. *Nat. Commun.* **2017**, *8*, 181. [CrossRef] [PubMed]
24. Blanksby, S.J.; Ellison, G.B. Bond dissociation energies of organic molecules. *Acc. Chem. Res.* **2003**, *36*, 255–263. [CrossRef] [PubMed]
25. Schawartz, J.; Hornyák, M.; Süts, T. Crystal structure of 5-amino-1H-1,2,3-triazole-4-carboxamide. *J. Chem. Soc. Perkin Trans.* **1974**, *2*, 1849.
26. Becke, A.D. Density-functional thermochemistry. III. The role of exact exchange. *J. Phys. Chem.* **1993**, *98*, 5648. [CrossRef]
27. Stephens, P.J.; Devlin, F.J.; Chabalowski, C.F.; Frisch, M.J. Ab Initio Calculation of Vibrational Absorption and Circular Dichroism Spectra Using Density Functional Force Fields. *J. Phys. Chem.* **1994**, *98*, 11623. [CrossRef]
28. Westwell, M.S.; Searle, M.S.; Williams, D.H. Empirical Correlations between Thermodynamic Properties and Intermolecular Forces. *J. Am. Chem. Soc.* **1995**, *117*, 5013. [CrossRef]

molecules

MDPI

Article

Synthesis and Properties of Energetic MOFs Based on Bis(3-Nitro-1*H*-1,2,4-triazole-5-yl) Amine: Advancing High Thermal Stability and Low Sensitivity

Shiluo Chen [1,2], Jinxin Wang [2], Yuteng Cao [2], Kangcai Wang [2], Haijun Yang [1,*] and Tianlin Liu [2,*]

1 School of Materials and Chemistry, Southwest University of Science and Technology, Mianyang 621010, China; cslswust@163.com
2 Institute of Chemical Materials, China Academy of Engineering Physics, Mianyang 621900, China; wangjinxin20@gscaep.ac.cn (J.W.); caoyuteng18@gscaep.ac.cn (Y.C.); wangkangcai@caep.cn (K.W.)
* Correspondence: yanghaijun@swust.edu.cn (H.Y.); hollandtian@caep.cn (T.L.)

Abstract: Energetic metal–organic frameworks (E-MOFs) have recently emerged as a promising strategy to address the long-standing challenge of reconciling energy and sensitivity in energetic materials. Nitrogen-rich compounds, with their abundant nitrogen atoms and superior enthalpy of formation, are particularly beneficial for forming multiple coordination bonds while simultaneously elevating the energy content. This makes them ideal ligand molecules for constructing E-MOFs. In this work, we report the synthesis and structural design of a novel series of E-MOFs, constructed from the nitrogen-rich energetic ligand BNTA and a range of alkali metals (Na–Rb, compounds **2**–**5**). The research indicates that the synthesized E-MOFs exhibit high thermal stability and low sensitivity. Specifically, Compound **3** displays a high decomposition temperature of 285 °C, with impact sensitivity and friction sensitivity values exceeding 40 J and 360 N, respectively. Moreover, Compound **3** also exhibits excellent computational detonation performance. Significantly, this study demonstrates how the aromatic character, coordination chemistry, and intermolecular interactions work synergistically to enhance the stability and safety of E-MOFs, thereby establishing fundamental criteria for engineering the next generation of energetic frameworks.

Keywords: E-MOFs; alkali metal; BNTA; high thermal stability; low sensitivity

1. Introduction

Balancing the inherent contradiction between energy and sensitivity is a crucial challenge in the field of energetic materials [1,2]. Achieving this balance can significantly enhance the comprehensive performance of energetic materials, thereby unlocking their potential for broader applications [3–5]. Metal–organic frameworks (MOFs) have emerged as highly promising candidates to address this challenge, owing to their exceptional structural tunability, remarkable chemical and thermal stability, and unparalleled design flexibility [6–8]. These attributes have attracted substantial research interest in recent years [9–12]. The synthesis of MOFs typically involves the straightforward formation of coordination complexes between metallic centers and polytopic organic connectors [13,14]. Energetic MOFs (E-MOFs), a specialized subclass of metal–organic frameworks, represent an emerging category of porous materials formed by integrating high-energy ligands with metal nodes [15]. These materials exhibit remarkable energy density, tunable architectures, and versatile characteristics, making them highly valuable for applications in propellant and explosive technologies [16]. By carefully selecting appropriate ligands, it is possible to construct energetic MOFs (E-MOFs) with diverse interesting structures [17–19]. These structures can effectively enhance the thermal stability and safety of the compounds, thereby mitigating the energy-sensitivity trade-off [20–22]. This approach not only offers a viable solution to the long-standing dilemma but also paves the way for the development of next-generation energetic materials with optimized performance [23,24].

It is worth noting that nitrogen-rich (N-rich) ligands, due to their high content of nitrogen and oxygen atoms, can form multiple coordination bonds with metal ions [25–28]. This leads to the formation of compounds with a high heat of formation (HOF), high density, enhanced thermal stability, good oxygen balance, and environmental friendliness [29–31]. As a result, the past decades have witnessed substantial progress in the development of energetic metal–organic frameworks (E-MOFs) employing N-rich coordinating ligands, such as triazoles, tetrazoles, and tetrazines [32–35]. As illustrated in Figure 1, continuous research efforts have led to the successful synthesis of a series of N-rich E-MOFs that exhibit remarkable thermal stability. E-MOFs incorporating higher nitrogen content and imino-bridged organic ligands tend to display superior thermal stability, as evidenced by their elevated decomposition temperatures [36–41]. For instance, potassium [5,5′-bitetrazole]-1,1′-diide (K_2BT) and potassium 4-(1H-tetrazole-5-ylamino)-1,2,4,5-tetrazine-1-one (K_2TATZO) exhibit notably high decomposition temperatures of 424 °C and 371.9 °C, respectively. Moreover, the abundance of nitrogen atoms in these ligands allows for extensive metal–ligand coordination, while the ligand–ligand interactions are primarily mediated by weaker forces [42–44]. This structural arrangement reduces mechanical sensitivity to external stimuli, thereby enhancing overall safety [45]. The above research findings demonstrate that E-MOFs based on nitrogen-rich energetic ligand molecules hold immense potential as candidates for high thermal stability and low sensitivity explosives [46–48].

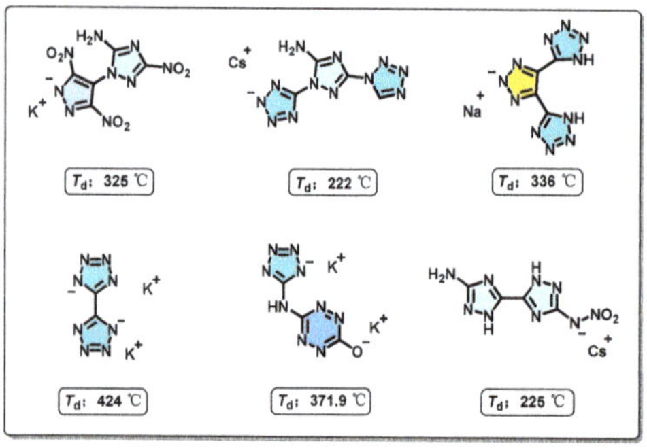

Figure 1. Some reported N-rich E-MOFs with high thermal stability.

In this study, BNTA was employed as an organic ligand due to its excellent thermal stability and low sensitivity to external stimuli. The combination with alkali metal cations provides a potential approach to develop E-MOFs with enhanced thermal stability and reduced sensitivity. Accordingly, four novel E-MOFs were constructed from alkali metals and the $BNTA^{2-}$ anion through simple preparation methods. The structures of compounds **2–5** were thoroughly characterized, and their physicochemical and energetic parameters were comprehensively evaluated using both experimental and theoretical approaches. These E-MOFs exhibit abundant coordination bonding interactions between the metal centers and energetic ligands, as well as significant hydrogen bonds and π–π interactions among the ligands themselves. These intermolecular interactions collectively contribute to the satisfactory thermal stability and mechanical insensitivity of the compounds. Moreover, compound **3** has the highest performance of energy, thermal stability, and mechanical sensitivity, endowing it with the greatest potential for practical applications.

2. Results and Discussion

2.1. Single-Crystal Structure

The crystals of compounds **3–5**, suitable for X-ray diffraction, were prepared by evaporation in aqueous solution. Compound **3** (CCDC 2454221) crystallizes in the triclinic $P-1$ space group, with each unit cell containing four asymmetric formula units (Z = 2). Notably, the calculated crystal density of this compound reaches a high value of 1.941 g·cm^{-3} at 295 K.

As shown in Figure 2a, the asymmetric unit comprises four K^+ (K1, K2, K3, and K4) ions, two $BNTA^{2-}$ anions, and four water molecules. The four K^+ ions exhibit distinct coordination environments with the surrounding $BNTA^{2-}$ anions and water molecules. Specifically, each K1 and K3 ion is connected to seven neighboring atoms. K1 is coordinated with four nitrogen atoms and three oxygen atoms, whereas K3 is coordinated with three nitrogen atoms and four oxygen atoms. The coordination modes of K2 and K4 are slightly different due to the presence of water molecules in their vicinity. Each K2 and K4 ion are connected to six and seven neighboring atoms, respectively. K2 is coordinated with one nitrogen atom and five oxygen atoms, while K4 is coordinated with one nitrogen atom and six oxygen atoms. The lengths of the K–O coordination bonds range from 2.710 to 3.315 Å, while the K–N coordination bonds span from 2.732 to 3.232 Å.

As a result of these diverse coordination forms, compound **3** exhibits a unique structural arrangement. Specifically, it forms ladder-like layer stackings, as shown in Figure 2b. This arrangement of anions and cations, with the alternating distribution of $BNTA^{2-}$ anions and coordinated K^+ ions within the same layer, facilitates interlayer sliding and compression. This structural feature effectively reduces mechanical sensitivities toward external stimuli, thereby enhancing the overall stability and safety of the compound.

To further investigate the stacking modes of compound **3**, as shown in Figure 2c, K^+ ions coordinate with adjacent $BNTA^{2-}$ anions and bridging water molecules via ligand–metal bonds, forming an extended one-dimensional chain-like structure with alternating K–O–K–O connectivity. Meanwhile, $BNTA^{2-}$ anions interact with K^+ ions through coordination bonds and are regularly distributed around the chain. To observe its three-dimensional structure (Figure 2d), the K^+ ions in the compound exhibit a hexagonal arrangement when observed from the a-axis direction. To further visually illustrate this phenomenon, by simplifying the N/O atoms around the K^+ ions, it is found that there is a regular hexagonal coordination geometry between K^+ ions, forming a honeycomb-like structure.

Compound **4** (CCDC 2454222) crystallizes in the *monoclinic* $P\,2_1/c$ space group, with each unit cell containing four asymmetric formula units (Z = 4). Remarkably, the calculated crystal density reaches up to 2.731 g·cm^{-3} at 293 K. As shown in Figure 3a, the asymmetric unit contained four Cs^+ (Cs1, Cs2, Cs3, and Cs4), two $DNTA^{2-}$ anions, and three water molecules. Four Cs^+ ions have different coordination forms with surrounding $BNTA^{2-}$ anions and water molecules. Each Cs1 and Cs3 is connected to the surrounding $BNTA^{2-}$ anion and water molecules through coordination bonds, interacting with eight neighboring atoms. Cs1 connects to four N atoms and four O atoms, while Cs3 connects to two N atoms and six O atoms. Meanwhile, the number of coordinating atoms of Cs2 and Cs4 is completely different. Cs2 connects to six N atoms and three O atoms, while Cs4 connects to one N atom and five O atoms. These Cs^+ ions, with distinct coordination modes, are regularly distributed in the holes formed by the alternating and staggered appearance of $BNTA^{2-}$ anions (Figure 3b). The lengths of Cs–O coordination bonds are in the range of 3.101–3.547 Å, yet Cs–N coordination bonds are from 3.102 to 3.760 Å. The staggered arrangement of anions and the regular distribution of cations can effectively release the mechanical sensitivity caused by external stimuli, ensuring the safety of the compound.

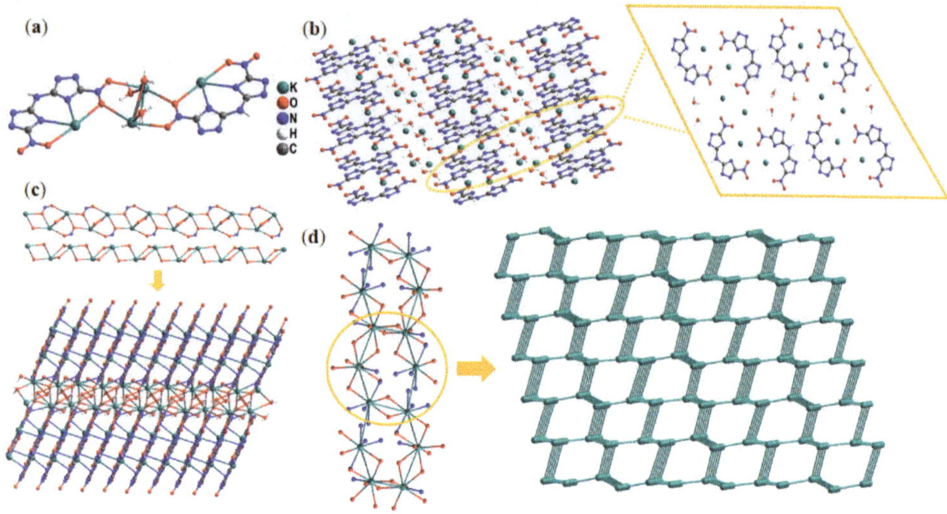

Figure 2. (**a**) Asymmetric unit of compound **3**. (**b**) The layered-like crystal packing of compound **3** and the molecular arrangement in the same layer. (**c**) The chain-like structure and layered structure extended via coordination and hydrogen bonding in compound **3**. (**d**) Simplified unrelated atoms and connected nodes and networks observed along the a-axis.

To gain insights into the complicated 3D network of compound **4**, we penetratingly dissected the coordination mode of the Cs^+ ions, $BNTA^{2-}$ ligands, and water molecules. As shown in Figure 3c, Cs^+ ions construct an intricate 3D network structure through coordination with $BNTA^{2-}$ anions and water molecules. This structural complexity arises from not only the face-to-face staggered arrangement of $BNTA^{2-}$ anions but also the additional perpendicular orientation of some $BNTA^{2-}$ anions relative to the plane formed by these staggered units. To more clearly visualize the 3D arrangement, the non-coordinating atoms in the $BNTA^{2-}$ anions were simplified (Figure 3d). It can be observed that the $BNTA^{2-}$ anions are distributed around the chain-like structures formed by the Cs^+ ions. When only the Cs^+ ions are considered, they form polygonal structures of varying shapes. These polygons alternate in arrangement, collectively constructing an intricate 3D network.

Compound **5** (CCDC 2454223) crystallizes in the *monoclinic* $P\,2_1/n$ space group, with each unit cell containing four asymmetric formula units (Z = 4). Remarkably, the calculated crystal density reaches up to 2.582 $g{\cdot}cm^{-3}$ at 150 K. As shown in Figure 4a, the asymmetric unit contained two Rb^+ (Rb1 and Rb2) and one $DNTA^{2-}$ anion. Two Rb^+ ions have different coordination forms with surrounding $BNTA^{2-}$ anions. The lengths of the Rb–O coordination bonds are in the range of 2.963–3.417 Å, yet Rb–N coordination bonds are from 2.981 to 3.527 Å. Both Rb1 and Rb2 are connected to the surrounding $BNTA^{2-}$ through coordination bonds, interacting with eight neighboring atoms. Rb1 connects to five N atoms and three O atoms, while Rb2 connects to three N atoms and five O atoms. Compound **5** exhibits the same $BNTA^{2-}$ anion arrangement as compound **3**, but due to the absence of water molecules in its crystal lattice, it displays better layered stacking. Rb^+ ions are uniformly and regularly distributed around these anions through coordination bonds (Figure 4b). Better-layered stacking, more conducive to intermolecular sliding and compression, endows compounds with better safety.

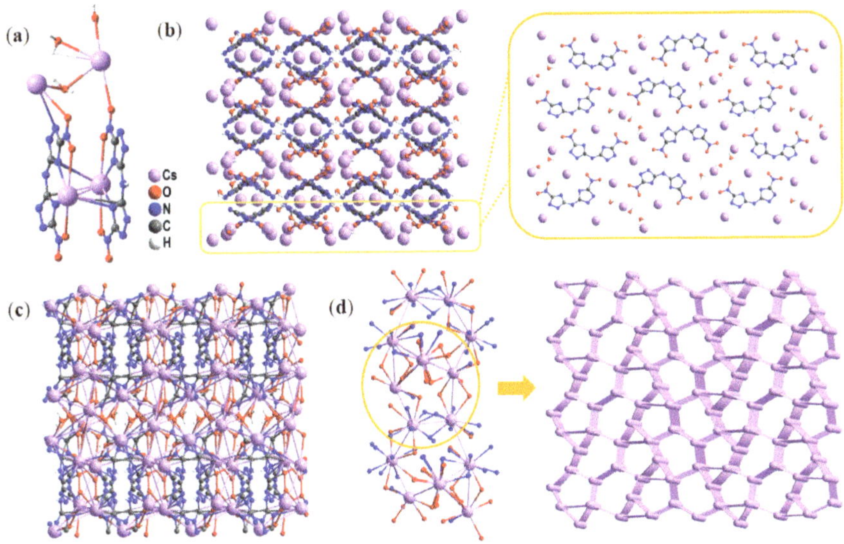

Figure 3. (**a**) Asymmetric unit of compound **4**. (**b**) The layered-like crystal packing of compound **4** and the molecular arrangement in the same layer. (**c**) Complex three-dimensional structure constructed by Cs^+ ions, $BNTA^{2-}$ anions, and water molecules together. (**d**) Simplified unrelated atoms and connected nodes and networks.

To further reveal the stacking forms in compound **5**, as shown in Figure 4c, a complex 3D network structure is formed between Rb^+ ions and planes constructed by misaligned $BNTA^{2-}$ anions. Compared to compound **4**, compound **5** has better intuition and regularity. The single-crystal structure of compound **5** is also simplified to give us insight into the intricate network. In Figure 4d, it can be observed that the Rb^+ centers are arranged in alternating triangular and quadrilateral configurations, and the structure overall exhibits an irregular arrangement form. Expanded structural illustrations better demonstrate this unique connectivity pattern. Further expansion can better observe this phenomenon.

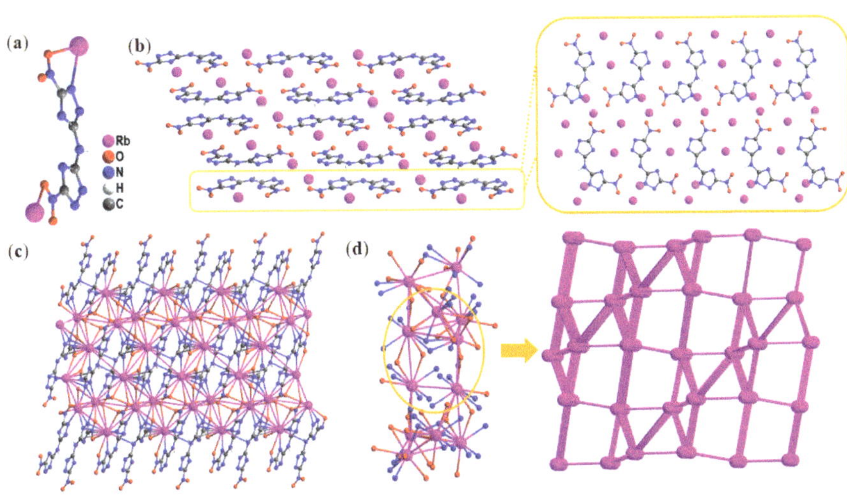

Figure 4. (**a**) Asymmetric unit of compound **5**. (**b**) The layered-like crystal packing of compound **5** and the molecular arrangement in the same layer. (**c**) Complex three-dimensional structure constructed by Rb^+ ions and $BNTA^{2-}$ anions. (**d**) Simplified unrelated atoms and connected nodes and networks.

2.2. Thermal Behaviors

Thermal stability is one of the significant indicators for evaluating the safety of energetic compounds, which is related to their potential applications under special conditions. Thermogravimetric analysis (TG) and differential scanning calorimetry (DSC) were applied to investigate the thermal stabilities of compounds **2–5** at 5 °C·min^{-1} in the N_2 atmosphere (Figure 4). Beforehand, compounds **2–5** were dried in air.

The thermal analysis results demonstrate that all synthesized compounds (**2–5**) exhibit excellent thermal stability, with decomposition temperatures ranging from 283 to 287 °C, which is higher than [Cs (ABTNA)H_2O] $_n$ [41]. For compound **2**, the DSC curve (Figure 5a) reveals two endothermic peaks at 85 °C and 116 °C, which correspond to weight loss observed in the TG curve. These peaks are attributed to the volatilization of free water and crystalline water within the compound. Subsequently, a pronounced exothermic peak is observed, with an onset temperature of 283 °C and a peak temperature of 299 °C.

A similar decomposition trend is observed for compound **3** (Figure 5b), which exhibits an endothermic peak at 181 °C, likely due to the evaporation of crystalline water. Compound **3** then displays a remarkably sharp exothermic peak, with an onset temperature of 285 °C and a peak temperature of 312 °C. In contrast, the decomposition curves of compounds **4** and **5** differ significantly from those of compounds **2** and **3**. The TG curves of compounds **4** and **5** each show only one distinct weight-loss step, while their DSC curves exhibit notable differences (Figure 5c,d). For compound **4**, a single sharp and intense exothermic peak is observed in the DSC curve, with an onset temperature of 287 °C and a peak temperature of 306 °C. Compound **5**, on the other hand, displays an endothermic peak at 239 °C without a corresponding mass loss in the TG curve, suggesting that this peak likely represents the melting of compound **5**. This is followed by a sharp exothermic peak with an onset temperature of 285 °C and a peak temperature of 308 °C. Overall, these results indicate that all compounds (**2–5**) possess good thermal stability, with high decomposition temperatures and distinct thermal behaviors that reflect their unique structural characteristics.

In addition, the TG-DSC analysis further reveals that compounds **2** and **3** undergo several stages of weight loss and a single stage of heat release, primarily associated with the collapse of the 3D framework and the thermal decomposition of the energetic components [38]. In contrast, compounds **4** and **5** each experience only one stage of weight loss and heat release, also accompanied by the collapse of the 3D framework and the thermal decomposition of the energetic components. From these observations, it can be concluded that the enhanced thermal stability of compounds **2–5** is partly attributed to the robustness of their framework structures, which are stabilized by multiple coordination bonds. Generally, a higher number of coordination bonds correlate with increased thermal stability, as these bonds provide additional structural reinforcement and resistance to thermal degradation.

2.3. Energetic and Safety Characteristics

Detonation characteristics and mechanical sensitivity are crucial indicators for evaluating the energy output and safety performance of energetic materials. Compounds **2–5** exhibited excellent densities, ranging from 1.860 to 2.733 g·cm^{-3}, which exceed [Na(H_2BTT)(H_2O)$_2$] $_n$ [38]. The enthalpy of formation (HOF) for these compounds was calculated using the Gaussian 09 software package, while their detonation performance was evaluated using the EXPLO5 (v6.02) code [49], based on the obtained HOF and density values. The enthalpy of formation for compounds **2–5** lies between −273.4 and −491.5 kJ·mol^{-1}. However, due to the computational limitations of EXPLO5 (v6.02), detonation performance calculations were only performed for compounds **2** and **3**. The results indicated that compound **3** exhibited superior detonation performance, with a detonation velocity of 8844 m·s^{-1} and a detonation pressure of 26.88 GPa. This enhanced performance is primarily attributed to compound 3′s more negative enthalpy of formation (-429.5 kJ·mol^{-1}) and higher density (1.941 g·cm^{-3} at room temperature). These calculation results demonstrate that compound **3** is a highly energetic compound with excellent detonation performance, surpassing traditional explosive molecules such as TNT (detonation velocity: 6881 m·s^{-1}; detonation pressure: 19.5 GPa) and RDX (detonation velocity: 8750 m·s^{-1}) [50,51].

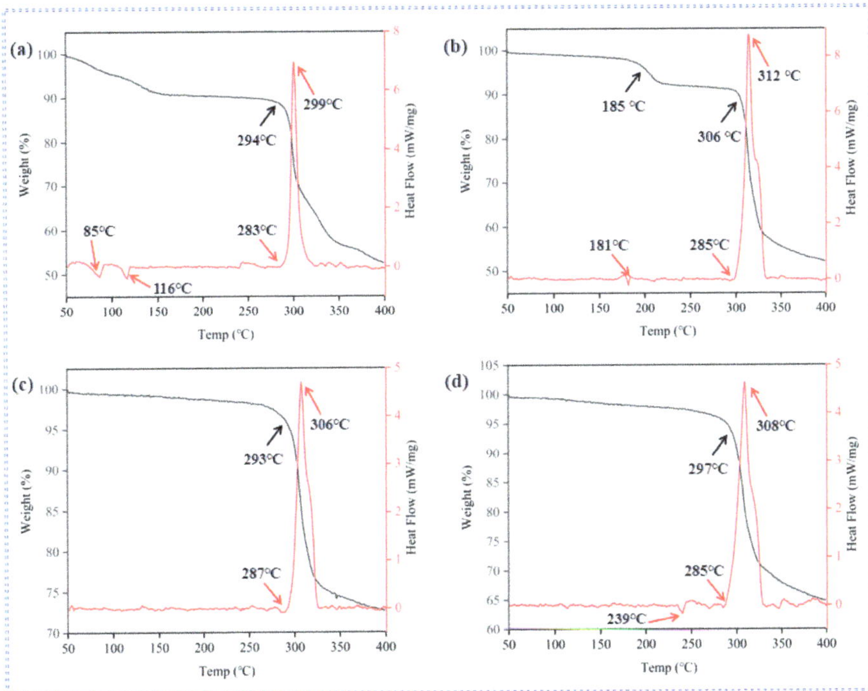

Figure 5. (**a**) TG-DSC curves of compound **2**. (**b**) TG-DSC curves of compound **3**. (**c**) TG-DSC curves of compound **4**. (**d**) TG-DSC curves of compound **5**.

The mechanical sensitivities of the energetic materials were assessed using the BAM standard sensitivity testing method. Prior to testing, compounds **2–5** were thoroughly dried to ensure accurate measurements. The experimental results revealed that compounds **2** and **3** exhibited impact sensitivities and friction sensitivities both exceeding 40 J and 360 N, respectively. In contrast, compounds **4** and **5** showed slightly lower sensitivity. Specifically, compound **4** had an impact sensitivity of 15 J and a friction sensitivity of 240 N, while compound **5** exhibited values of 20 J and 288 N, respectively. Despite these minor differences, all four target compounds qualify as low-sensitivity energetic materials, fully meeting the stringent application requirements. Notably, compounds **2** and **3** demonstrated superior mechanical insensitivity, which can be partly attributed to their robust hydrogen-bonding interactions and well-ordered three-dimensional structures. These structural features effectively dampen the mechanical sensitivity induced by external stimuli, thereby enhancing overall safety and stability. Overall, although slight variations in mechanical sensitivity were observed among the four compounds, they all outperformed typical energetic molecules such as RDX [51] and HMX [52] are both in Table 1, highlighting their potential as advanced energetic materials.

Table 1. Physicochemical and energetic properties of compounds **2–5** in comparison with traditional explosives TNT, RDX, and HMX.

Compd.	T_d [a] [°C]	ρ [b] [g·cm^{-3}]	$\Delta_f H$ [c] [kJ·mol^{-1}]	D [d] [m·s^{-1}]	P [e] [GPa]	IS [f] [J]	FS [g] [N]
2	283	1.860	−273.4	7116	17.51	>40	>360
3	285	1.941	−429.5	8844	26.88	>40	>360
4	287	2.756	−485.0	-	-	15	240
5	285	2.582/2.556 [h]	−491.5	-	-	20	288
[Na(H_2BTT)(H_2O)$_2$]$_n$ [38]	336	1.706	-	8120	22.83	>40	>360
[Cs (ABTNA) H_2O]$_n$ [41]	225	2.413	974.05	6780	23.9	60	360
TNT [50]	295	1.65	−67.0	6881	19.5	15	353
RDX [51]	205	1.81	86.3	8750	34.2	7.5	120
HMX [52]	279	1.90	116.1	9144	41.5	7	112

[a] temperature of decomposition; [b] density measured at room temperature; [c] calculated molar enthalpy of formation; [d] calculated detonation velocity; [e] calculated detonation pressure; [f] impact sensitivity; [g] friction sensitivity; [h] density at 298.15 K converted by formula $\rho_{(298.15K)} = \rho - 0.188 \times (298.15 - T)/1000$.

2.4. Hirshfeld Surface

To gain a deeper understanding of the correlations between the physicochemical properties and intermolecular interactions of compounds **3–5**, Hirshfeld surface analysis, 2D fingerprint spectra, and individual atomic interaction proportions were conducted [53]. In compound **3**, the widely distributed red regions on the Hirshfeld surface (Figure 6a,c) represent strong coordination interactions between K–O and K–N atoms, as well as hydrogen bonding interactions. In contrast, blue and white spots indicate weaker π–π contacts between different layers of the molecule [54,55]. These close contacts are primarily attributed to nitrogen atoms on the 1,2,4-triazole ring and oxygen atoms from nitro groups or water molecules. Consequently, N–H and O–H hydrogen bonding interactions dominate the crystal structure, accounting for up to 40.5%. Coordination interactions between K–O and K–N atoms contribute an additional 19.1%. Additionally, N–N and N–O contacts, which are closely related to π–π interactions, account for 7.4% and 8.1%, respectively.

Compound **4** exhibits a similar pattern of weak non-covalent interactions to compound **3**. As shown in Figure 6d,f, hydrogen bonding and coordination bonds play a major role in its crystal structure. N–H and O–H hydrogen bonding interactions account for 35.2%, while K–O and K–N coordination interactions reach 29.3%. In contrast, compound **5** lacks water molecules, and coordination bonds are the primary driving force in its structure rather than hydrogen bonding (Figure 6g,i). Rb–N and Rb–O coordination bonds dominate the crystal structure of compound **5**, accounting for over 48.0%, while hydrogen bonding interactions contribute only 7.8%. Additionally, π–π stacking interactions mediated by N–N, O–O, and N–O contacts are present, contributing 7.8%, 8.9%, and 11.5%, respectively. This phenomenon is primarily attributed to the more parallel alignment of anionic planes in compound **5**.

The 2D fingerprint spectrum provides a visual representation of the distribution of weak interactions in these compounds. In all three compounds (**3–5**), two prominent spikes are observed, corresponding to high proportions of O–H and N–H hydrogen bonding (Figure 6b,e,h). The distribution of this hydrogen bonding is like some rich nitrogen E-MOFs, such as potassium 5-(hydrazine carbonyl)-3,4-dinitropyrazole [45]. The presence of these hydrogen bonds can effectively improve the stability of compounds and reduce their mechanical sensitivity to external stimuli. The high nitrogen and oxygen content in these compounds enhances molecular density while also reducing mechanical sensitivity. However, the abundance of coordination bonds and π–π interactions significantly contributes to the insensitivity of compounds **3–5** to external mechanical stimuli. The synergistic modulation of multiple weak interactions, including but not limited to hydrogen bonding, π–π stacking, and coordination bonding, represents a proven and versatile strategy for engineering E-MOFs with simultaneously enhanced thermal stability and reduced mechanical sensitivity.

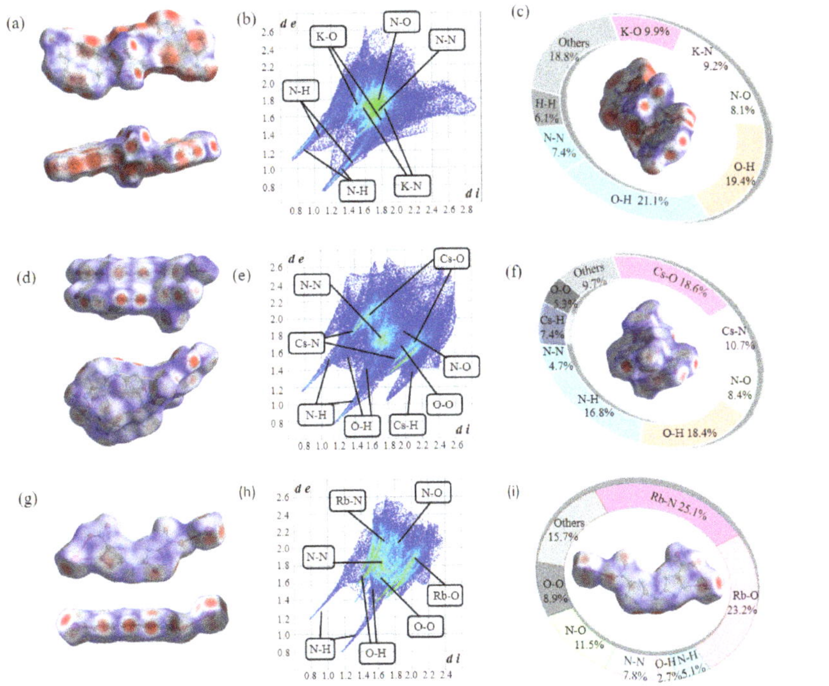

Figure 6. (**a**) Hirshfeld surface of compound **3**. (**b**) The 2D fingerprint plot of compound **3**. (**c**) The related individual atomic interaction proportion of compound **3**. (**d**) The Hirshfeld surface of compound **4**. (**e**) The 2D fingerprint plot of compound **4**. (**f**) The related individual atomic interaction proportion of compound **4**. (**g**) The Hirshfeld surface of compound **5**. (**h**) The 2D fingerprint plot of compound **5**. (**i**) The related individual atomic interaction proportion of compound **5**.

3. Methods

The reagents and drugs used in this work were purchased from Aladdin Reagent Company (Shanghai, China). The synthetic route for compounds **2–5** is depicted in Figure S1. The starting material [56,57], Bis(3-nitro-1*H*-1,2,4-triazol-5-yl) amine (1), was synthesized based on our previous work. Utilizing compound **1** as the precursor, four E-MOFs were successfully obtained via a metathesis reaction [58,59]. Herein, the preparation method for compound **2** is described in detail. In a typical experiment, an ethanol solution of BNTA was first prepared. Sodium carbonate and water were then sequentially added to this solution. The reaction mixture was subsequently heated to 85 °C and maintained at this temperature for two hours. During the heating process, a substantial amount of precipitate formed. This precipitate was filtered, washed, and dried to yield a brown-yellow solid powder. These samples were characterized by ^{1}H NMR, ^{13}C NMR (400 AVANCE, Bruker, Biospin, AG, Ettlingen, Germany), IR (Spectrum II), single-crystal X-ray diffraction (D8 Venture, Bruker, AXS, Karlsruhe, Germany), and powder X-ray diffraction, etc. The details for all compounds (**2–5**) are provided in the Supplementary Materials.

4. Conclusions

In summary, a series of alkali metal-based energetic metal–organic frameworks (E-MOFs) were rapidly and conveniently synthesized using nitrogen-rich imino-bridged bis(1,2,4-triazole) as the ligand. All compounds (**2–5**) were thoroughly characterized. Structural studies of these E-MOFs revealed diverse coordination modes and three-dimensional architectures. These compounds feature not only abundant coordination bonds and hydrogen bonds but also significant π–π interactions, as evidenced by Hirshfeld surface analysis. These intermolecular interactions collectively contribute to their mechanical insensitivity. In terms of properties, compounds **2–5** exhibit high densities, ranging from 1.860 to 2.756 $g\cdot cm^{-3}$. Additionally, they possess high thermal decomposition temperatures and excellent mechanical stability against external stimuli. Notably, compound 3 stands out with a density of 1.941 $g\cdot cm^{-3}$, a high decomposition temperature of 285 °C, and low mechanical sensitivity (impact sensitivity >40 J; friction sensitivity >360 N). It also demonstrates remarkable detonation performance, with a detonation velocity of 8844 $m\cdot s^{-1}$ and a detonation pressure of 26.88 GPa, surpassing that of the traditional explosive RDX. These experimental and computational results clearly demonstrate that constructing E-MOFs using nitrogen-rich energetic molecules is an effective strategy to simultaneously enhance thermal stability and energy performance while reducing mechanical sensitivity to external stimuli. This approach paves the way for the development of next-generation energetic materials with optimized performance.

Supplementary Materials: The following supporting information can be downloaded at: https://www.mdpi.com/article/10.3390/molecules30122478/s1, Experimental details, potential hazards, NMR spectra, mass spectra, IR spectra, crystallographic data, and theoretical calculations, etc. Refs. [60,61] are cited in Supplementary Materials.

Author Contributions: T.L. and H.Y. took the duty of oversight and leadership of Molecules and the responsibility for the research activity planning and execution, including mentorship external to the core team; K.W. mainly participated in article guidance and technical support; S.C. mainly designed the experimental structure and formulation or evolution of overarching research goals and aims; J.W. and Y.C. made significant contributions to crystal analysis in articles, such as crystal analysis, crystal optimization, and crystal mapping. All authors have read and agreed to the published version of the manuscript.

Funding: This work was supported by the National Natural Science Foundation of China (No. 22175161).

Institutional Review Board Statement: Not applicable.

Informed Consent Statement: Not applicable.

Data Availability Statement: CCDC 2454221–2454223 contain the supplementary crystallographic data for this paper. These data can be obtained free of charge via www.ccdc.cam.ac.uk/data_request/cif (accessed on 27 April 2025), or by emailing data_request@ccdc.cam.ac.uk, or by contacting.

Conflicts of Interest: The authors declare no conflict of interest.

References

1. Barton, L.M.; Edwards, J.T.; Johnson, E.C.; Bukowski, E.J.; Sausa, R.C.; Byrd, E.F.C.; Orlicki, J.A.; Sabatini, J.J.; Baran, P.S. Impact of Stereo- and Regiochemistry on Energetic Materials. *J. Am. Chem. Soc.* **2019**, *141*, 12531–12535. [CrossRef] [PubMed]
2. Xue, Z.H.; Huang, B.B.; Li, H.Z.; Yan, Q.L. Nitramine-Based Energetic Cocrystals with Improved Stability and Controlled Reactivity. *Cryst. Growth Des.* **2020**, *20*, 8124–8147. [CrossRef]
3. Tariq, Q.U.N.; Manzoor, S.; Ling, X.; Dong, W.S.; Lu, Z.J.; Wang, T.W.; Xu, M.Q.; Younis, M.A.; Yu, Q.Y.; Zhang, J.G. Fabrication, Characterization, and Performance Evaluation of Thermally Stable [5,6]-Fused Bicyclic Energetic Materials. *ACS Appl. Mater. Interfaces* **2024**, *16*, 52613–52623. [CrossRef] [PubMed]

4. Zhang, J.Y.; Fei, T.; Meng, J.W.; Cai, J.X.; Zhang, L.; Pang, S.P.; He, C.L. Taming of Trinitromethyl-Oxadiazole to Access High Density and High Oxygen Balance via a Dual Modulation Strategy. *Def. Technol.* **2025**, *43*, 142–149.
5. Li, Y.; Jiang, Z.Y.; Liu, Z.J.; Li, B.D. Constructing Single- or Dual-Layer Biomass Composite Energetic Material through Self-Assembly of Biomass Polyphenol Structural Materials. *Langmuir* **2024**, *40*, 12313–12321. [CrossRef]
6. Imchen, S.; Amrit, P. Metal–Organic Framework (MOF)-based Single Atom Catalysts (SACs) for Photocatalytic Energy Conversion. *ChemistrySelect* **2024**, *9*, e202403516. [CrossRef]
7. Wang, C.; Liu, D.; Lin, W.B. Metal–organic frameworks as a tunable platform for designing functional molecular materials. *J. Am. Chem. Soc.* **2013**, *135*, 13222–13234. [CrossRef]
8. Howarth, A.J.; Liu, Y.; Li, P.; Li, Z.; Wang, T.C.; Hupp, J.T.; Farha, O.K. Chemical, thermal and mechanical stabilities of metal–organic frameworks. *Nat. Rev. Mater.* **2016**, *1*, 15018. [CrossRef]
9. Gebremariam, S.K.; Varghese, A.M.; Ehrling, S.; Wahedi, Y.A.; Hajaj, A.A.; Dumée, L.F.; Karanikolos, G.N. Hierarchically Porous Structured Adsorbents with Ultrahigh Metal–Organic Framework Loading for CO_2 Capture. *ACS Appl. Mater. Interfaces* **2024**, *16*, 50758–50799. [CrossRef]
10. Devendran, A.; Shahmirzaee, M.; Nagai, A. Recent Progress on Metal Organic Framework and Covalent Organic Framework Based Solid-State Electrolyte Membranes for Lithium Battery Applications. *ChemElectroChem* **2024**, *11*, e202300847. [CrossRef]
11. Wang, S.; Wang, Q.Y.; Feng, X.; Wang, B.; Yang, L. Explosives in the Cage: Metal-Organic Frameworks for High-Energy Materials Sensing and Desensitization. *Adv. Mater.* **2017**, *29*, 1701898. [CrossRef] [PubMed]
12. Check, B.; Bairley, K.; Santarelli, J.; Pham, H.T.B.; Park, J. Applications of Electrically Conductive Metal–Organic Frameworks: From Design to Fabrication. *ACS Mater. Lett.* **2025**, *7*, 465–488. [CrossRef]
13. Ye, B.C.; Li, W.H.; Zhang, X.; Chen, J.; Gao, Y.; Wang, D.S.; Pan, H.G. Advancing Heterogeneous Organic Synthesis with Coordination Chemistry-Empowered Single-Atom Catalysts. *Adv. Mater.* **2024**, *36*, 2402747. [CrossRef] [PubMed]
14. Jerozal, R.T.; Pitt, T.A.; MacMillan, S.N.; Milner, P.J. High-concentration self-assembly of zirconium-and hafnium-based metal–organic materials. *J. Am. Chem. Soc.* **2023**, *145*, 13273–13283. [CrossRef]
15. Zhang, S.; Yang, Q.; Liu, X.; Qu, X.; Wei, Q.; Chen, S.P.; Gao, S.L. High-energy metal–organic frameworks (HE-MOFs): Synthesis, structure and energetic performance. *Coord. Chem. Rev.* **2016**, *307*, 292–312. [CrossRef]
16. Tan, B.J.; Dou, J.K.; Yang, X.; Li, W.J.; Zhang, J.; Zhang, P.F.; Mo, H.C.; Lu, X.M.; Wang, B.Z.; Liu, N. Application and prospects of EMOFs in the fields of explosives and propellants. *Dalton Trans.* **2024**, *53*, 13308–13319. [CrossRef]
17. Xu, L.P.; Qiao, J.; Xu, S.Y.; Zhao, X.Y.; Gong, W.J.; Huang, T.Z. Constructing Strategies and Applications of Nitrogen-Rich Energetic Metal—Organic Framework Materials. *Catalysts* **2020**, *10*, 690. [CrossRef]
18. Cao, Y.T.; Liu, Y.; Zhang, W.Q. Pentazolate Anion: A Rare and Preferred Five-Membered Ligand for Constructing Pentasil-Zeolite Topology Architectures. *Angew. Chem. Int. Ed.* **2024**, *63*, e202317355. [CrossRef] [PubMed]
19. Zhang, W.Q.; Wang, K.C.; Li, J.C.; Lin, Z.E.; Song, S.W.; Huang, S.L.; Liu, Y.; Nie, F.D.; Zhang, Q.H. Stabilization of the Pentazolate Anion in a Zeolitic Architecture with $Na_{20}N_{60}$ and $Na_{24}N_{60}$ Nanocages. *Angew. Chem. Int. Ed.* **2018**, *57*, 2592–2595. [CrossRef]
20. Gao, H.X.; Zhang, Q.H.; Shreeve, J.N.M. Fused Heterocycle-Based Energetic Materials (2012–2019). *J. Mater. Chem. A* **2020**, *8*, 4193–4216. [CrossRef]
21. Sher, F.; Hayward, A.; Guerraf, E.A.; Ziani, I.; Hrnjić, H.; Boškailo, E.; Nemțanu, M.R. Advanced metal-organic frameworks for superior carbon capture, high-performance energy storage and environmental photocatalysis—A critical review. *J. Mater. Chem. A* **2024**, *12*, 27932–27973. [CrossRef]
22. Lin, J.D.; Chen, F.; Xu, J.G.; Zheng, F.K.; Wen, N. Framework-Interpenetrated Nitrogen-Rich Zn (II) Metal-Organic Frameworks for Energetic Materials. *ACS Appl. Nano. Mater.* **2019**, *2*, 5116–5124. [CrossRef]
23. Xu, R.; Yan, Z.Z.; Yang, L.; Wang, Q.Y.; Tong, W.C.; Song, N.M.; Han, J.M.; Zhao, Y. Nanoscale Homogeneous Energetic Copper Azides@Porous Carbon Hybrid with Reduced Sensitivity and High Ignition Ability. *ACS Appl. Mater. Interfaces* **2018**, *10*, 22545–22551. [CrossRef]
24. Liu, G.R.; Bu, R.P.; Huang, X.; Zhong, K.; Jiao, F.B.; Wei, S.H.; Li, H.Z.; Zhang, C.Y. Energetic Cocrystallization as the Most Significant Crystal Engineering Way to Create New Energetic Materials. *Cryst. Growth Des.* **2022**, *22*, 954–970. [CrossRef]
25. Cao, S.N.; Ma, X.F.; Ma, X.H.; Cen, P.P.; Wu, Y.W.; Yang, J.H.; Liu, X.Y.; Xie, G.; Chen, S.P. Modulating energetic performance through decorating the nitrogen-rich ligands in high-energy MOFs. *Dalton Trans.* **2020**, *49*, 2300–2307. [CrossRef]
26. Xu, J.G.; Sun, C.; Zhang, M.J.; Liu, B.W.; Li, X.Z.; Liu, J.; Wang, S.H.; Zheng, F.K.; Guo, G.C. Coordination Polymerization of Metal Azides and Powerful Nitrogen-Rich Ligand toward Primary Explosives with Excellent Energetic Performances. *Chem. Mater.* **2017**, *29*, 9725–9733. [CrossRef]
27. Xu, Z.; Hou, T.Y.; Zhang, X.P.; Yang, F.; Zhang, L.N.; Jiang, S.J.; Lu, M.; Xu, Y.G. Self-Assembly and Controllable Regulation of 1,5′-Bitetrazolate-2 N-oxide-Based EMOFs toward High Dimensionality. *Cryst. Growth Des.* **2024**, *24*, 461–470. [CrossRef]
28. Ma, Q.; Huang, S.L.; Lu, H.C.; Nie, F.D.; Liao, L.Y.; Fan, G.J.; Huang, J.L. Energetic Cocrystal, Ionic Salt, and Coordination Polymer of a Perchlorate Free High Energy Density Oxidizer: Influence of pKa Modulation on Their Formation. *Cryst. Growth Des.* **2019**, *19*, 714–723. [CrossRef]

29. Li, F.G.; Zhao, W.Y.; Chen, S.T.; Zhang, T.L.; Zhou, Z.N.; Yang, L. Nitrogen-Rich Alkali Metal Salts (Na and K) of [Bis (N, N-bis(1*H*-tetrazol-5-yl) amine)-zinc (II)] Anion: Syntheses, Crystal Structures, and Energetic Properties. *Z. Anorg. Allg. Chem.* **2015**, *641*, 911–916. [CrossRef]
30. Banik, B.K.; Sahoo, B.M.; Kumar, B.V.V.R.; Panda, K.C.; Jena, J.; Mahapatra, M.K.; Borah, P. Green synthetic approach: An efficient eco-friendly tool for synthesis of biologically active oxadiazole derivatives. *Molecules* **2021**, *26*, 1163. [CrossRef]
31. Liu, Q.Q.; Jin, B.; Zhang, Q.C.; Shang, Y.; Guo, Z.C.; Tan, B.S.; Peng, R.F. Nitrogen-Rich Energetic Metal-Organic Framework: Synthesis, Structure, Properties, and Thermal Behaviors of Pb (II) Complex Based on N, N-Bis(1*H*-tetrazole-5-yl)-Amine. *Materials* **2016**, *9*, 681. [CrossRef]
32. Zhang, G.F.; Hao, X.; Zou, Y.B.; Liu, S.C.; Wei, J.J.; Dong, Z.; Ye, Z.W. Towards Advanced N-rich Energetic Explosives: Based on Tetrazole and Triazole Groups with Large Conjugated Systems and Extensive Hydrogen Bonds. *J. Mater. Chem. A* **2024**, *12*, 33249–33256. [CrossRef]
33. Aromi, G.; Barrios, L.A.; Roubeau, O.; Gamez, P. Triazoles and Tetrazoles: Prime Ligands to Generate Remarkable Coordination Materials. *Coord. Chem. Rev.* **2011**, *255*, 485–546. [CrossRef]
34. Myers, T.W.; Brown, K.E.; Chavez, D.E.; Scharff, R.J.; Veauthuer, J.M. Laser Initiation of Fe (II) Complexes of 4-Nitro-pyrazolyl Substituted Tetrazine Ligands. *Inorg. Chem.* **2017**, *56*, 2297–2303. [CrossRef]
35. Stetsiuk, O.; Alexandre, A.; Narcis, A. 1,2,4,5-Tetrazine based ligands and complexes. *Dalton Trans.* **2020**, *49*, 5759–5777. [CrossRef]
36. Li, C.; Zhang, M.; Chen, Q.S.; Li, Y.Y.; Gao, H.Q.; Fu, W.; Zhou, Z.M. 1-(3,5-Dinitro-1*H*-pyrazol-4-yl)-3-nitro-1*H*-1,2,4-triazol-5-amine (HCPT) and its Energetic Salts: Highly Thermally Stable Energetic Materials with High-performance. *Dalton Trans.* **2016**, *45*, 17956–17965. [CrossRef] [PubMed]
37. Zhang, G.F.; Hao, X.; Zou, Y.B.; Liu, S.C.; Wei, J.J.; Dong, Z.; Ye, Z.W. Several energetic MOFs based on the N-rich energetic materials and alkali metals: Towards high detonation performances and good stabilities. *CrystEngComm* **2025**, *27*, 277–283. [CrossRef]
38. Chen, D.; Jing, D.; Zhang, Q.; Xue, X.G.; Gou, S.H.; Li, H.Z.; Nie, F.D. Study of Six Green Insensitive High Energetic Coordination Polymers Based on Alkali/Alkali-Earth Metals and 4,5-Bis(tetrazol-5-yl)-2*H*-1,2,3-triazole. *Chem. Asian J.* **2017**, *40*, 3141–3149. [CrossRef]
39. Finger, L.H.; Schröder, F.G.; Sundermeyer, J. Synthesis and Characterisation of 5,5′-Bistetrazolate Salts with Alkali Metal, Ammonium and Imidazolium Cations. *Z. Anorg. Allg. Chem.* **2013**, *639*, 1140–1152. [CrossRef]
40. Zhang, C.; Guo, Y.; Xiao, L.B.; Gao, H.X.; Zhao, F.Q.; Huang, J.; Lv, X.Q.; Ma, H.X. 4-(1*H*-Tetrazole-5-yl-Amino)-1,2,4,5-Tetrazin-1-one (TATzO) Metal Salts: Promising Pyrotechnic Agent. *J. Therm. Anal.* **2024**, *149*, 2697–2706. [CrossRef]
41. Wang, T.W.; Zhou, J.Y.; Zhang, Q.; Zhang, L.; Zhu, S.G.; Li, Y. Novel 3D cesium(i)-based EMOFs of nitrogen-rich triazole derivatives as "green" orange-light pyrotechnics. *New J. Chem.* **2020**, *44*, 1278–1284. [CrossRef]
42. Rajak, R.; Kumar, N.; Ghule, V.D.; Dharavath, S. Highly Dense N–N-Bridged Dinitramino Bistriazole-Based 3D Metal–Organic Frameworks with Balanced Outstanding Energetic Performance. *ACS Appl. Mater. Interfaces* **2024**, *16*, 20670–20680. [CrossRef] [PubMed]
43. Zhang, X.P.; Hou, T.Y.; Wu, Y.G.; Xu, Z.; Lin, Q.H.; Wang, P.C.; Xu, Y.G.; Lu, M. Crowding-out Strategy to Enhance Energetics of 3-nitro-1,2,4-triazol-5-one based E-MOFs at the Molecular Level through Adjusting pH. *Chem. Eng. J.* **2023**, *466*, 143091. [CrossRef]
44. Zhang, L.; Dong, W.S.; Lu, Z.J.; Wang, T.W.; Zhang, C.; Zhou, Z.N.; Zhang, J.G. Synthesis and Characterization of Thermally Stable Energetic Complexes with 3,5-diaminopyrazolone-4-oxime as a Nitrogen-rich Ligand. *CrystEngComm* **2022**, *24*, 5519–5526. [CrossRef]
45. Cao, Y.T.; Wang, K.C.; Song, S.W.; Liu, Y.; Zhang, W.Q. Fabrication of Energetic Metal–Organic Frameworks: Potassium 5-carboxylato-3,4-dinitropyrazole and Potassium 5-(hydrazine carbonyl)-3,4-dinitropyrazole. *Inorg. Chem.* **2023**, *62*, 17199–17206. [CrossRef] [PubMed]
46. Li, W.; Wang, K.C.; Qi, X.J.; Jin, Y.H.; Zhang, Q.H. Construction of a Thermally Stable and Highly Energetic Metal—Organic Framework as Lead-free Primary Explosives. *Cryst. Growth Des.* **2018**, *18*, 1896–1902. [CrossRef]
47. Rajak, R.; Kumar, P.; Dharavath, S. Mixed-Metallic Energetic Metal–Organic Framework: New Structure Motif for Potential Heat-Resistant Energetic Materials. *Cryst. Growth Des.* **2024**, *24*, 2142–2148. [CrossRef]
48. Zhang, L.F.; Wang, Y.; Wang, Y.F.; Liu, S.; Zhang, N.; Yang, M.M.; Ma, H.X.; Guo, Z.Q. N-Methylene-C bridged Tetrazole and 1,2,4-triazole Energetic Salts as Promising Primary Explosives. *CrystEngComm* **2024**, *26*, 143–152. [CrossRef]
49. Sućeska, M. *EXPLO5 V6.05.02[CP]*; Brodarski Institute: Zagreb, Croatia, 2019.
50. Dacons, J.C.; Sitzmann, M.E. Synthesis of 2,4,6-trinitrophenyl Derivatives of Heterocyclic Compounds. *J. Heterocycl. Chem.* **1977**, *14*, 1151–1155. [CrossRef]
51. Bian, C.M.; Feng, W.J.; Lei, Q.Y.; Huang, H.F.; Li, X.; Wang, J.L.; Li, C.; Xiao, Z.L. A Facile Synthesis of Energetic Salts Based on Fully Nitroamino-Functionalized [1,2,4] triazolo [4,3-b] [1,2,4] triazole. *Dalton Trans.* **2020**, *49*, 368–374. [CrossRef]
52. Brill, T.B.; Gongwer, P.E.; Williams, G.K. Thermal decomposition of energetic materials. 66. Kinetic compensation effects in HMX, RDX, and NTO. *J. Phys. Chem.* **1994**, *98*, 12242–12247. [CrossRef]

53. Spackman, M.A.; Jayatilaka, D. Hirshfeld Surface Analysis. *CrystEngComm* **2009**, *11*, 19–32. [CrossRef]
54. Li, X.; Sun, Q.; Lin, Q.H.; Lu, M. [N-N=N-N]-Linked Fused Triazoles with π-π Stacking and Hydrogen Bonds: Towards Thermally Stable, Insensitive, and Highly Energetic Materials. *Chem. Eng. J.* **2021**, *406*, 126817. [CrossRef]
55. Cao, Y.T.; Cai, Z.W.; Shi, J.H.; Zhang, Q.H.; Liu, Y.; Zhang, W.Q. A Heat-Resistant and Insensitive Energetic Material Based on the Pyrazolo-Triazine Framework. *Energy Mater. Front.* **2022**, *3*, 26–31. [CrossRef]
56. Zhou, Z.M.; Li, C.; Deng, C.L.; Wang, M. Structure Preparation Method and Performances of High-Energy Heat-Resistant Explosive CPTY. CN201711400638A, 7 September 2018.
57. John, W.F.; Robert, D.C.; Richard, D.G. Facile Entry into the 3*H*,9*H*-Bis[1,2,4] Triazolo-[1,5-*a*:5′,1′-*d*] [1,3,5]Triazinium (5/6/5 Tricyclic NNN) System. *Tetrahedron Lett.* **2006**, *47*, 7707–7709.
58. Li, Y.N.; Wang, B.; Chang, P.; Hu, J.J.; Chen, T.; Wang, Y.L.; Wang, B.Z. Novel Catenated N_6 Energetic Compounds Based on Substituted 1,2,4-triazoles: Synthesis, Structures and Properties. *RSC Adv.* **2018**, *8*, 13755–13763. [CrossRef]
59. Dippold, A.A.; Klapötke, T.M.; Winter, N. Insensitive Nitrogen-Rich Energetic Compounds Based on the 5,5′-Dinitro-3,3′-bi-1,2,4-triazol-2-ide Anion. *Eur. J. Inorg. Chem.* **2012**, *2012*, 3474–3484. [CrossRef]
60. Frisch, M.J.; Trucks, G.; Schlegel, H.B.; Scuseria, G.E. *Gaussian 09, Revision D. 01*; Gaussian Inc.: Wallingford, CT, USA, 2009.
61. Jenkins, H.D.B.; Tudela, D.; Glasser, L. Lattice Potential Energy Estimation for Complex Ionic Salts from Density Measurements. *Inorg. Chem.* **2002**, *41*, 2364–2367. [CrossRef]

molecules

MDPI

Article

DFT Study on Fused *N*-Heteroaromatic Frameworks: Stability, Aromaticity, and Energetic Insights from Five-Membered Fused Six-Membered *N*-Heteroaromatic Skeletons

Zujia Lu [1], Cong Li [1], Shaoqun Li [1,2], Qiyao Yu [1,*] and Jianguo Zhang [1,*]

1 State Key Laboratory of Explosion Science and Technology, Beijing Institute of Technology, Beijing 100081, China; 15201657211@163.com (Z.L.); 18656815056@163.com (C.L.); 3220235040@bit.edu.cn (S.L.)
2 State Key Laboratory of Transient Chemical Effects and Control, Shaanxi Applied Physics and Chemistry Research Institute, Xi'an 710061, China
* Correspondence: qiyaoyu@bit.edu.cn (Q.Y.); zjgbit@bit.edu.cn (J.Z.)

Abstract: The five-membered fused six-membered nitrogen heteroaromatic ring system is a crucial skeleton in the design and synthesis of energetic compounds. Based on this skeleton, many high-performance energetic compounds have been synthesized. However, to date, no one has conducted a systematic study on the characteristics of this skeleton itself. To assess how the number and position of nitrogen atoms affect the energy and stability of this type of skeleton, one to four nitrogen-substituted skeleton molecules were analyzed using Density Functional Theory (DFT) calculations. Natural population analysis (NPA), Laplacian bond order (LBO) analysis, aromaticity studies, and enthalpy of formation calculations were performed. Patterns observed in the computational results were summarized, and their potential correlations were analyzed. Based on these findings, design recommendations for derivatives of these skeletons in energetic compounds were proposed to serve as a reference for energetic material chemists.

Keywords: energetic skeleton; DFT; wave function analysis; molecule screening

1. Introduction

Energetic materials, with their high energy density, rapid energy release rate, and portability, have become indispensable in the military and aerospace fields. They also have significant applications in civilian fields such as civil engineering and mining. Since the time of Alfred Bernhard Nobel, generations of energetic material scientists have been dedicated to developing materials with superior performance. The energetic derivatization of nitrogen heteroaromatic rings is currently the leading approach in designing and synthesizing new energetic compounds [1,2]. This trend stems from the balanced energy and stability of nitrogen heteroaromatic rings. Compared to pure carbon skeletons, the C-N, N-N, and N=N bonds in nitrogen heteroaromatic rings release more energy during decomposition and recombination, while their aromaticity provides additional stability. Notable energetic modifications of nitrogen heteroaromatic rings include NTO (nitrotriazolone) [3] and TNP (3,4,5-trinitropyrazole) [4] based on azole ring skeletons, and LLM-105 (2,6-diamino-3,5-dinitropyrazine-1-oxide) [5] and ICM-102 (2,4,6-triamino-5-nitropyrimidine-1,3-dioxide) [6] based on azine ring skeletons. These explosive molecules are designed and synthesized based on single nitrogen heteroaromatic rings.

Subsequently, energetic material chemists have expanded their focus to a broader chemical space involving bridged and fused nitrogen heteroaromatic rings, achieving

https://doi.org/10.3390/molecules30051101

significant outcomes. Energetic compounds based on fused nitrogen heteroaromatic rings typically possess high density and structural stability, making them ideal for designing heat-resistant explosives and high-energy insensitive explosives. In this field, Professors Yang [7,8] and Tang [9] from Nanjing University of Science and Technology and Pang [10] from Beijing Institute of Technology have made outstanding contributions.

Among fused skeletons, the five-membered fused six-membered nitrogen heteroaromatic ring skeleton is the most common due to its simplicity in synthesis and structural stability. The number of such skeletons is relatively limited. We found that there are only 213 five-membered fused six-membered aromatic ring skeletons with one to four nitrogen atoms, making a comprehensive and systematic study of this skeleton both feasible and necessary, particularly to investigate the impact of different numbers and positions of nitrogen atoms on the skeleton's energy and stability.

In this study, we use Density Functional Theory (DFT) to perform natural population analysis [11] (NPA), Laplacian bond order [12,13] (LBO) analysis, aromaticity studies (localized orbital locator function for π-electrons [14,15] (LOL-π), multicenter bond order [16] (MCBO)), enthalpy of formation (EOF) calculations, and correlation studies of related properties. We summarize the relevant patterns, propose design suggestions for energetic compounds based on these skeletons, and aim to provide a reference for energetic material chemists.

2. Result and Discussion

2.1. NPA Charges Analysis

As shown in Figure 1, the nitrogen atoms in FR213 can be classified into three types. The first type consists of nitrogen atoms on C-N bonds shared by fused rings, labeled as N_S. The second type includes nitrogen atoms that form N-H bonds, labeled as N_H. The third type comprises nitrogen atoms where the bonds at both ends are single and double bonds, labeled as N_0. The lone pairs of the first two types of nitrogen atoms (N_S and N_H) participate in the conjugation of the ring system, with their electron-donating conjugation effect being greater than their electron-withdrawing inductive effect. The lone pairs of the third type of nitrogen atom (N_0) are perpendicular to the π orbitals and do not participate in ring system conjugation, exhibiting electron-withdrawing inductive and conjugation effects. In organic chemistry, N_S and N_H are referred to as pyrrole-like nitrogen atoms, while N_0 is referred to as pyridine-like nitrogen atoms. It is important to note that the requirement to maintain the aromaticity of the skeleton molecule stipulates that N_S and N_H are mutually exclusive. That is, if a skeleton contains N_S, it cannot contain N_H, and vice versa. Their total number can only be 1.

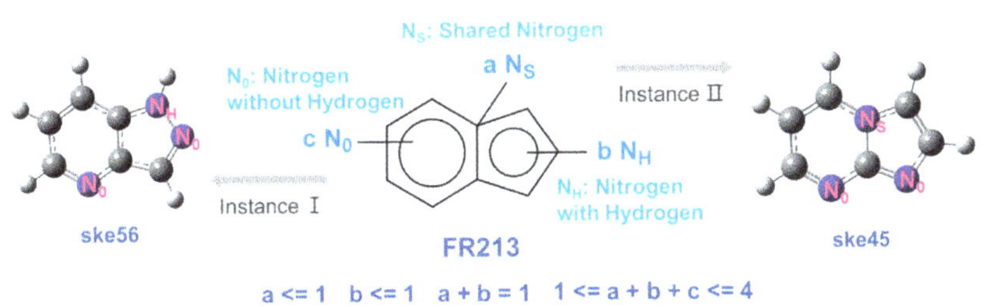

Figure 1. Schematic diagram of three types of nitrogen atoms in FR213; a, b, and c represent the number of N_S, N_H, and N_0 atoms, respectively.

Natural population analysis (NPA) [11] was conducted on FR213, and the population diagrams with specific charge values for each skeleton are shown in Figure S2, with the corresponding color mapping provided in Figure S3. To investigate the relationship between NPA charge distribution and skeleton structure, the most negative and most positive NPA charges of FR213, along with their corresponding atomic types, are visualized in Figure 2. It can be observed that no significant differences are found between the skeletons with varying numbers of substituted nitrogen atoms.

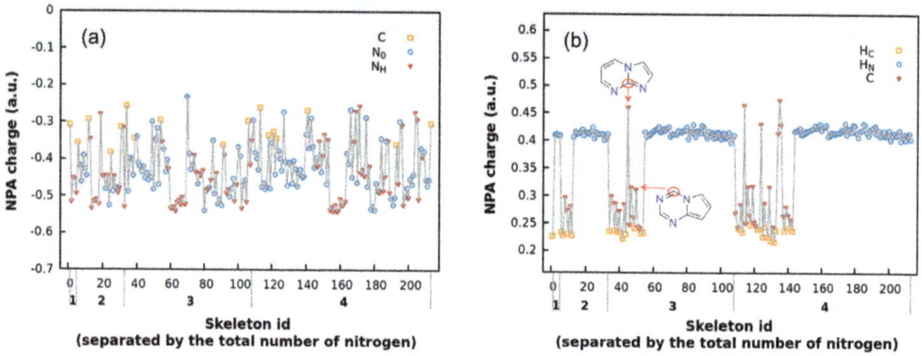

Figure 2. The most negative (**a**) and positive (**b**) NPA charge values of each skeleton.

As shown in Figure 2a, the most negative charges correspond to three atomic types, C, N_0, and N_H, with the general trend of corresponding charge values being $N_H < N_0 < C$. Combining this with the structural analysis of each skeleton, we find that the atomic types associated with the most negative charges are determined by the following patterns:

(1) Electronegativity trend: N > C > H.
(2) N_H tends to carry more negative charge due to the electron-donating effect of the H atom.
(3) Catenated nitrogen substructures (N3 or N4) reduce the negative charge on nitrogen atoms compared to isolated nitrogen atoms. Nitrogen atoms at the ends of catenated nitrogen substructures carry more negative charge, whereas those in the middle exhibit less.
(4) Both shared nitrogen (N_S) and carbon (C_S) atoms exhibit less negative charge compared to their isolated counterparts of the same type.
(5) For the same type of atom, those in five-membered rings carry more negative charge than those in six-membered rings.
(6) Charges tend to alternate in distribution; that is, if an adjacent atom carries a higher positive charge, the atom in question is likely to carry more negative charge.

Using the patterns described above, the atom corresponding to the most negative charge in a framework can be roughly identified. For example, if a framework molecule contains an N_H group on a five-membered ring that is not part of a consecutive nitrogen substructure, it can typically be identified as the atom with the most negative charge. Similarly, nitrogen atoms on a five-membered ring near a shared bond often carry the most negative charge, as they are in close proximity to C_S atoms, which carry more positive charge.

As shown in Figure 2b, most of the atoms with the highest positive charge are hydrogen atoms attached to nitrogen (H_N), with a smaller number being carbon (C) or hydrogen atoms attached to carbon (H_C). In fact, whenever an N-H bond is present, the atom with the highest positive charge will always be H_N. In the absence of an N-H bond, the atom with the highest positive charge could be either H_C or a carbon atom located between two or three nitrogen atoms. The patterns used to determine the most negatively charged atom can also assist in identifying the most positively charged atom. For instance, when no N-H bond is present, a C_S atom located between two nitrogen atoms can typically be identified as the atom with the highest positive charge.

2.2. Laplacian Bond Order Analysis

Laplacian bond order (LBO), proposed by Lu [12] and particularly suitable for studying covalent bonding, has been demonstrated to exhibit a strong linear correlation with bond dissociation energies [13] (BDE), which is advantageous for analyzing cyclic systems where calculating BDEs is challenging. Therefore, we primarily use LBO to assess the structural stability of FR213. As depicted in Figure 3, there is a general trend where the average minimum skeleton LBO ($LBO_{min\text{-}noH}$) gradually decreases as the number of nitrogen atoms in the skeleton increases. This trend becomes more pronounced when nitrogen atoms are present in six-membered rings. Six-membered rings, being six-center six-electron systems, exhibit lower average electron density compared to the five-center six-electron systems of five-membered rings. The introduction of electron-withdrawing pyridine-like nitrogen intensifies the electron deficiency effect, further reducing the system's stability.

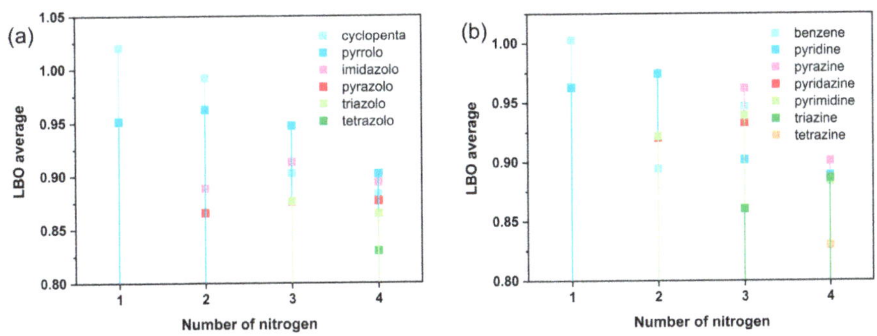

Figure 3. FR213 includes different numbers of nitrogen atoms subdivided by ring type for average $LBO_{min\text{-}noH}$; (**a**) subdivided by five-membered ring type; (**b**) subdivided by six-membered ring type.

As shown in Figure S4, the bond corresponding to the minimum LBO (LBO_{min}) is most often the N-H bond when N_H is present. Excluding N-H or C-H bonds and considering only the bonds within the ring skeleton, the LBO values generally follow the order C-C > C-N > N-N, which aligns with chemical knowledge. However, when a molecule's Lewis structure contains a double N-N bond (i.e., N=N), the LBO of the N-N bond can exceed that of C-N bonds. Additionally, when N_S is present, the smallest C-N bond is always the shared C-N bond, likely because the connected carbon atom has relatively fewer charges.

Excluding N-H and C-H bonds, the average $LBO_{min\text{-}noH}$ of skeletons with N_H is greater than those without N_H. This is due to the electron-donating inductive effect of hydrogen in N-H bonds, which strengthens the C-N or N-N bonds involving the nitrogen in the N-H bond. Moreover, the C-N bond in N_S skeletons often becomes the weakest bond, with its LBO being smaller than other C-N bonds. Therefore, in terms of bond stability within the ring skeleton, excluding N-H bonds and focusing only on $LBO_{min\text{-}noH}$, skeletons with N_H have an advantage. However, N_H-type skeletons include weaker N-H bonds, potentially leading to proton transfer and making the entire ring unstable.

To investigate the effect of the relative position of nitrogen atoms on bond strengths, we conducted a statistical analysis comparing the LBO of skeletons that include and exclude catenated nitrogen substructures. First, we examined the average LBO of the weakest N-N bonds (average $LBO_{min\text{-}N\text{-}N}$) in three nitrogen-substituted skeletons, considering both cases with and without the N3 substructure. It was found that when the N3 substructure was present, the average $LBO_{min\text{-}N\text{-}N}$ was 0.912, while without the N3 substructure, it was 0.914. Similarly, no statistically significant differences were observed in the average weakest LBO for other bond types. Next, we analyzed four nitrogen-substituted frameworks. When the N3 substructure was included, no statistically significant differences were observed in the average weakest LBO for various bond types. However, when the N_4 substructure was included, the average $LBO_{min\text{-}N\text{-}N}$ was 0.868, compared to 0.898 when the N_4 substructure was absent, showing a noticeable difference. For other bond types, no statistically significant differences were found.

We observed that the LBO of C-H bonds is not significantly affected by the number or position of substituted nitrogen atoms, and their average value of 0.872 can be considered a benchmark. When a framework with the in-ring bonds whose LBO is greater than 0.872 decomposes, it tends to lose protons first rather than undergo ring opening and can thus be considered a relatively stable structure. Figure 4 shows 12 skeleton molecules that meet this criterion, which we regard as ideal skeletons for energetic materials from a structural stability perspective. It should be noted that molecules containing N_H were not considered in this selection.

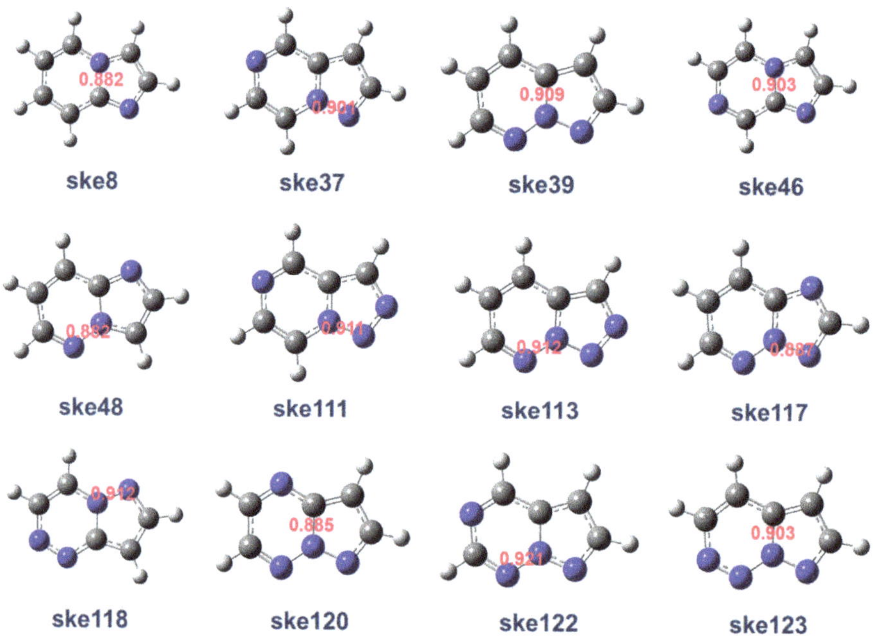

Figure 4. The top 12 skeletons sorted by $LBO_{min\text{-}noH}$ in FR213 (blue = N, gray = C, white = H).

2.3. LOL-π Analysis

To qualitatively assess the aromaticity of FR213, we used the localized orbital locator function for π electrons (LOL-π) and generated color-filled plots. These plots were created using the Multiwfn program, taking the LOL-π values at 1 Å above the plane of the skeleton. The results are shown in Figure S5. The blue-green-red color scale corresponds to LOL-π values in the range of 0 to 0.8.

It can be observed that the LOL-π paths traverse the entire skeleton in all molecules, indicating varying degrees of aromaticity. Certain bond regions display deeper red colors, indicating areas with higher concentrations of π electrons. These regions correspond to the double-bond areas in the Lewis structures of the skeleton molecules. Since all molecules in FR213 have only one resonance structure, this single Lewis structure effectively reflects the distribution of bond strength. Double-bond regions contain more π bond components, corresponding to richer π electron densities.

2.4. Multicenter Bond Order Analysis

The multicenter bond order (MCBO) can be used to quantitatively analyze the aromaticity of FR213, and normalized MCBO (NMCBO) can compare the aromaticity of rings of different sizes. As shown in Table 1, the presence of N_H atom leads to a decrease in the average NMCBO-5R (where NMCBO-5R refers to the NMCBO of five-membered rings; similarly, NMCBO-6R refers to the NMCBO of six-membered rings). N_H-6R decreases both the average NMCBO-6R and NMCBO-5R, while N_H-5R significantly increases the average NMCBO-6R. This indicates that N_S is beneficial for the aromaticity of five-membered rings, N_H-5R enhances the aromaticity of six-membered rings, and N_H-6R is detrimental to the aromaticity of six-membered rings. Generally, NMCBO-5R and NMCBO-6R both tend to decrease with an increasing number of nitrogen atoms in the six-membered ring. However, when N_H-5R is present, NMCBO-6R generally increases and no longer follows this trend. By analyzing these patterns, we can infer that the placement and type of nitrogen atoms in the fused ring systems significantly influence the aromaticity and stability of the compounds, as measured by NMCBO values.

Table 1. The average NMCBO for different types of six-membered rings within various skeleton types.

Skeleton Types	Six-Membered Rings	Average NMCBO-5R	Average NMCBO-6R
N_H-5R	benzene	0.51567	0.57977
	pyridine	0.51199	0.58495
	pyrazine	0.50957	0.58365
	pyridazine	0.50939	0.58466
	pyrimidine	0.50873	0.58265
	triazine	0.50859	0.58706
	Summary Average	**0.51087**	**0.58411**
N_H-6R	pyridine	0.52345	0.51568
	pyridazine	0.51281	0.51010
	pyrimidine	0.50961	0.50912
	pyrazine	0.49399	0.51652
	triazine	0.48808	0.50332
	tetrazine	0.46313	0.49413
	Summary Average	**0.50449**	**0.50922**
N_S	pyrazine	0.54020	0.53373
	pyridazine	0.53737	0.52897
	pyridine	0.53629	0.53414
	triazine	0.52938	0.52845
	pyrimidine	0.52827	0.53232
	tetrazine	0.52249	0.52450
	Summary Average	**0.53163**	**0.53039**

2.5. Enthalpy of Formation Analysis

The enthalpy of formation for FR213 was calculated using the definition method (for calculation details, see the SI). As shown in Figure 5, the overall trend indicates that the enthalpy of formation for the skeleton increases with increasing nitrogen content (Figure 5a), which is consistent with chemical intuition. Similarly, the density of the skeletons shows a comparable pattern (Figure S6).

We performed a multiple linear regression of the number of each type of bond in the skeleton against the enthalpy of formation, resulting in the fitting formula shown in Equation (1). Here, B_{CN} represents the number of C-N bonds, B_{NN} represents the number of N-N bonds, and B_{CC} represents the number of C-C bonds. Since all bonds in the ring skeletons are aromatic, we do not strictly distinguish between single and double bonds. The regression shows a very good correlation, with a correlation coefficient $R^2 = 0.98521$ and a root mean square error (RMSE) of 40.4 kJ mol^{-1}. Figure 5b compares the predicted values from the fitting formula with the actual values.

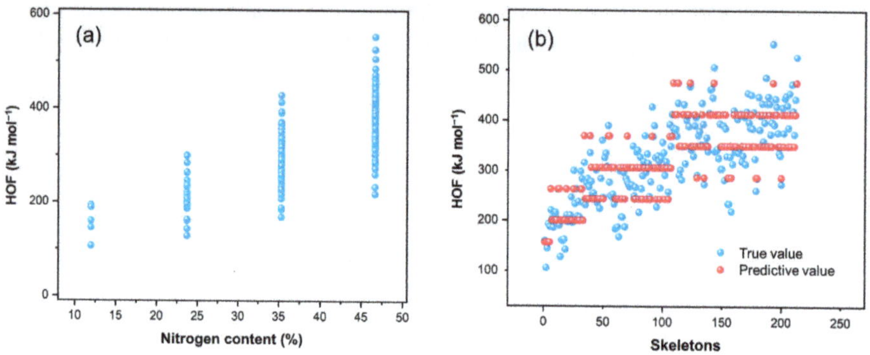

Figure 5. (**a**) Distribution of enthalpy of formation for FR213 with different nitrogen content; (**b**) scatter plot comparing predicted values from Equation (1) with true values.

From the coefficients of each term in Equation (1), we can infer the relative contribution of each type of bond to the enthalpy of formation. The order of contribution is N-N > C-N > C-C, with the coefficient for C-N bonds being three times that of C-C bonds and the coefficient for N-N bonds being ten times that of C-C bonds. This highlights the significant contribution of nitrogen-involved chemical bonds to the enthalpy of formation.

$$HOF = 33.04B_{CN} + 117.72B_{NN} + 11.31B_{CC} \quad (1)$$

In tetra-nitrogenous skeletons, when the number of N-N bonds is two, comparing two different arrangements of nitrogen atoms—either forming a direct N-N bond with N3 or having two separate N-N bonds—the corresponding average molecular enthalpies of formation are 404.9 and 421.6 kJ mol^{-1}, respectively. This indicates that under the same nitrogen content and N-N bond count, the linear chain structure of nitrogen atoms does not always exhibit higher energy than expected. Additionally, when comparing the average enthalpies of formation in tetra-nitrogenous skeletons with and without N-H bonds, they are 379.6 and 379.0 kJ mol^{-1}, respectively. This suggests that N-H bonds have minimal impact on energy levels.

We found that nitrogen atoms in six-membered rings (N_6), five-membered rings (N_5), and shared nitrogen atoms (N_S) contribute differently to the EOF of skeleton molecules. N_6 contributes the most, N_5 the least, and NS falls in between. Figure 6 shows the histogram of average EOF for skeletons composed of different nitrogen atom types. The numbers at the top of each bar represent the nitrogen atom composition: the hundreds digit indicates the number of N_6 atoms, the tens digit represents the number of N_S atoms, and the ones digit corresponds to the number of N_5 atoms. It can be observed that the average EOF generally increases with the composition number, supporting our findings. However, there are some exceptions to this trend, which can be attributed to certain nitrogen atom compositions containing more N-N bonds, thereby releasing additional energy. In the LBO analysis, we noted that an increase in N_6 relative to N_5 leads to a more significant decrease in stability. This is consistent with the fact that N_6 provides higher energy, offering an explanation for this behavior.

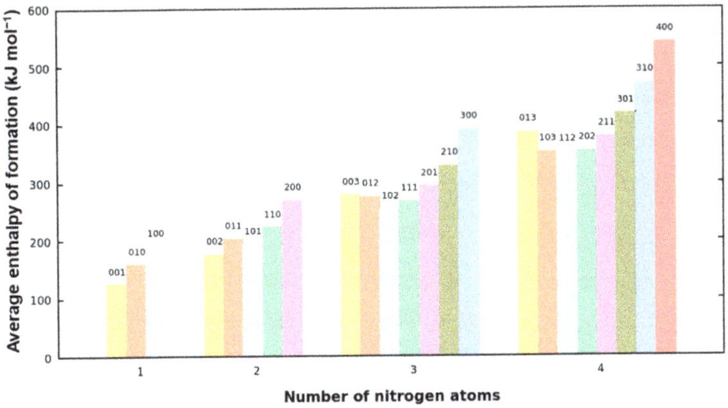

Figure 6. Average EOF of skeletons composed of different nitrogen atom types. The numbers at the top of each bar represent the nitrogen atom composition, with the hundreds digit indicating the number of N_6 atoms, the tens digit representing the number of N_S atoms, and the ones digit corresponding to the number of N_5 atoms. The data are grouped by the total number of nitrogen atoms.

2.6. Correlation Analysis

To explore the potential correlations among the structural features, aromaticity, stability, and HOMO-LUMO gap of FR213, we conducted a Pearson correlation analysis with a significance level of 0.05. As shown in Figure 7, N_H is strongly negatively correlated with the minimum LBO value, indicating that when N_H is present, the weakest bond in the molecule tends to be the N-H bond. Moreover, N_H shows moderate positive correlations with $LBO_{min\text{-}noH}$, $LBO_{ave\text{-}noH}$, and LBO_{ave} of the skeleton rings, suggesting its enhancing effect on other bonds within the skeleton. These findings align with the conclusions from our LBO analysis in Section 2.2.

Additionally, we observed that N_H is negatively correlated with the MCBO of the five-membered rings, particularly on six-membered rings, where it exhibits a more pronounced negative correlation with MCBO. This indicates that N_H on six-membered rings simultaneously reduces the aromaticity of five- and six-membered rings. Interestingly, N_H on five-membered rings shows a positive correlation with the MCBO of six-membered rings.

Furthermore, the MCBO of six-membered rings shows a weak negative correlation with the enthalpy of formation (EOF), implying that the enhanced aromaticity of six-membered rings slightly lowers the molecule's energy. Additionally, there is a significant positive correlation between the MCBO of six-membered rings and the HOMO-LUMO gap, suggesting that the aromaticity of six-membered rings has a greater impact on the overall stability of the molecule. $LBO_{min\_C\text{-}C}$ shows a significant negative correlation with NH-6R (correlation coefficient of −0.80), while it exhibits moderate positive correlations with MCBO-5R and MCBO-6R (correlation coefficients of 0.43 and 0.46, respectively). The underlying mechanism may be that NH-6R reduces the aromaticity of the skeleton, thereby indirectly weakening the C-C bond.

Therefore, from the perspective of enhancing the aromatic stability of the skeleton molecule, it is advisable to minimize the introduction of N_H on six-membered rings when designing energetic materials.

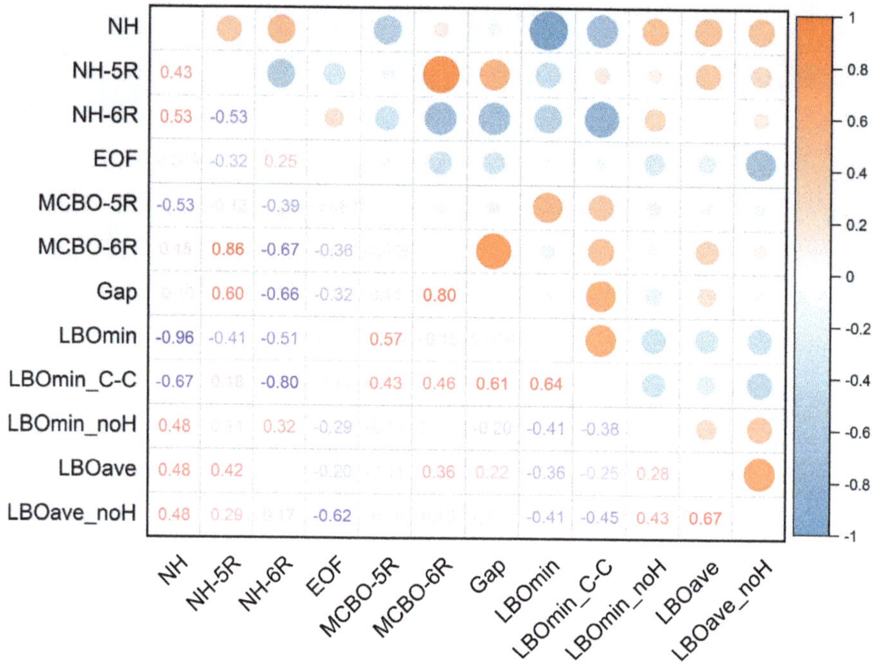

Figure 7. Pearson correlation heatmap of FR213-related properties. NH: N_H atom; NH-5R: N_H atom located in a five-membered ring; NH-6R: N_H atom located in a six-membered ring; EOF: Enthalpy of formation; MCBO-5R: MCBO of a five-membered ring; MCBO-6R: MCBO of a six-membered ring; Gap: HOMO-LUMO gap; LBOmin: Minimum LBO among all bonds; LBOmin_C-C: Minimum LBO of the C-C bond; LBOmin_noH: Minimum LBO of all bonds excluding C-H and N-H bonds; LBOave: Average LBO of all bonds; LBOave_noH: Average LBO of all bonds excluding C-H and N-H bonds.

3. Computing Method

Using SMILES encoding [17], we enumerated and deduplicated one to four nitrogen-containing five-membered fused six-membered skeletons with a self-developed program. We then used the Open Babel 3.1.1 program [18] to remove skeletons that did not satisfy aromaticity criteria, ultimately obtaining 213 five-membered fused six-membered nitrogen heteroaromatic ring skeletons. For ease of discussion, we name this system FR213 (where FR stands for Fused Rings), and the 213 skeletons are labeled with a numerical sequence prefixed by "ske". Their 2D structures are shown in Figure S1. The 3D structures were geometrically optimized and subjected to vibrational analysis at the M062X/def2TZVP level using the Gaussian 09 program [19]. All subsequent analyses are based on the optimized structures. NPA charges were calculated using the NBO module of Gaussian 09, while LBO, LOL-π, and MCBO analyses were performed using the Multiwfn 3.8 program [20,21].

4. Summary and Recommendations

Nitrogen, as an element with properties distinctly different from carbon, when introduced into five-membered and six-membered aromatic carbon frameworks, can cause significant changes in the properties of the original framework. The number and positions of nitrogen atoms can lead to subtle differences in the reactivity, stability, and energy of the molecular frameworks. Through systematic calculations and analyses of all 213 five-membered and six-membered aromatic nitrogen-containing frameworks

(FR213), ranging from mono-nitrogen to tetra-nitrogen substitutions, we have derived some general patterns.

(1) The nitrogen atoms in FR213 can be classified into three types, N_S, N_H, and N_0, as described in the main text. They exhibit completely different conjugative and inductive properties. Different atomic types and their relative positions can have subtle effects on the NPA charge population of the skeletons.
(2) In the context of LBO analysis, the LBO_{min} decreases as the number of nitrogen atoms increases, suggesting a reduction in the stability of the framework with higher nitrogen content. This trend is particularly pronounced when nitrogen atoms are incorporated into six-membered rings. N-H bonds are unequivocally the weakest bonds in the framework, yet their presence enhances the strength of other bonds in the framework. The LBO values of the bonds in the framework rings generally follow the trend C-C > C-N ≈ N=N > N-N. Among the C-N bonds, C-N_S is the weakest. Besides, the N3 substructure does not have a significant effect on bond strength, but the N4 substructure does result in relatively weaker N-N bonds.
(3) All molecules in FR213 have unique Lewis structures, and parameters such as bond length, LBO, and LOL-π paths can be associated with these structures. For example, regions corresponding to double bonds in the structure will exhibit shorter bond lengths, higher LBO values, and greater π electron density. Therefore, the strength of bonds can be qualitatively discussed based on the Lewis structures of the skeleton molecules.
(4) In the context of MCBO analysis, the presence and position of N-H have a significant impact on the aromaticity of the framework molecules. N-H on the five-membered ring enhances the aromaticity of the six-membered ring, while N-H on the six-membered ring weakens the aromaticity of both the five-membered and six-membered rings. The MCBO of the six-membered ring shows a clear positive correlation with the HOMO-LUMO gap.
(5) The EOF of FR213 is strongly correlated with the composition of its chemical bonds. By fitting the EOF to the quantities of C-C, C-N, and N-N bonds, we obtain a fitting equation with $R^2 = 0.98521$ and RMSE = 40.4 kJ mol^{-1}. The coefficients of this equation reflect the relative contributions of each bond type to the EOF. Specifically, the coefficient for the C-N bond is three times that of C-C, and the coefficient for the N-N bond is ten times that of C-C, indicating a significant contribution of nitrogen-containing bonds to the EOF. Furthermore, for a given bond composition, longer nitrogen chains do not result in a higher EOF. The presence or absence of N-H bonds has minimal impact on the EOF of the framework. Additionally, for the contribution to EOF: $N_6 > N_S > N_5$.

Based on these patterns, we can make some recommendations for designing energetic compounds based on such frameworks:

(1) In the synthesis of elemental explosives, skeletons containing N-H bonds are undesirable. Although N-H bonds can strengthen the framework, the N-H bond itself is the most unstable component, potentially leading to instability in explosive molecules, manifested as acidity or high sensitivity.
(2) Achieving a balance between energy and stability depends on the distribution of nitrogen atom positions. While there is a general trend of decreasing stability with increasing nitrogen content, as shown in Figure 4, many tri-nitrogen and tetra-nitrogen frameworks can still maintain good structural stability. Specifically, placing more than two nitrogen atoms on six-membered rings should be avoided, as nitrogen atoms on these rings have a more detrimental effect on the framework's stability. Moreover, continuous nitrogen chain structures do not confer energetic benefits and should be avoided in the framework to mitigate their negative impact on stability.

Overall, this research advances our understanding of nitrogen-containing aromatic frameworks and provides a foundational basis for the rational design of new energetic materials. By elucidating the relationships between structure, aromaticity, stability, and energetic properties, this work offers valuable insights for future synthetic and computational strategies aimed at optimizing the performance and stability of energetic compounds.

Supplementary Materials: The following supporting information can be downloaded at https://www.mdpi.com/article/10.3390/molecules30051101/s1; Figure S1: The 2D structures of FR213, along with their enthalpy of formation (in kJ mol^{-1}), are shown below; Figure S2: The geometric structure of optimized FR213 with NPA annotation; Figure S3: NPA charge color mapping. Red represents negative values, and green represents positive values; Figure S4: The LBO visualization of FR213, where red dots represent the bonds with LBO_{min}, and yellow dots represent the bonds with the $LBO_{min\text{-}noH}$; Figure S5: The LOL-π color-filled plots of FR213; Figure S6: Distribution of densities for FR213 with different nitrogen content. Ref. [22] is cited in the Supplementary Materials.

Author Contributions: J.Z. conceived the project. J.Z. and Z.L. designed this study and completed the main theoretical calculation work and data collection. C.L. and S.L. participated in data collection and analysis. Q.Y. checked the data for our manuscript and assisted us in completing it. All authors have read and agreed to the published version of the manuscript.

Funding: This research was funded by National Natural Science Foundation of China grant number 22175025.

Institutional Review Board Statement: Not applicable.

Informed Consent Statement: Not applicable.

Data Availability Statement: Data are contained within the article and Supplementary Materials.

Acknowledgments: We are thankful to the projects of NSFC (22175025) for their generous financial support.

Conflicts of Interest: The authors declare no conflict of interest.

References

1. Gao, H.; Shreeve, J.M. Azole-Based Energetic Salts. *Chem. Rev.* **2011**, *111*, 7377–7436. [CrossRef]
2. Zhou, J.; Zhang, J.; Wang, B.; Qiu, L.; Xu, R.; Sheremetev, A.B. Recent Synthetic Efforts towards High Energy Density Materials: How to Design High-Performance Energetic Structures? *FirePhysChem* **2022**, *2*, 83–139. [CrossRef]
3. Becuwe, A.; Delclos, A. Low-Sensitivity Explosive Compounds for Low Vulnerability Warheads. *Propellants Explos. Pyrotech.* **1993**, *18*, 1–10. [CrossRef]
4. Hervé, G.; Roussel, C.; Graindorge, H. Selective Preparation of 3,4,5-Trinitro-1 *H*-Pyrazole: A Stable All-Carbon-Nitrated Arene. *Angew. Chem. Int. Ed.* **2010**, *49*, 3177–3181. [CrossRef] [PubMed]
5. Pagoria, P.; Mitchell, A.; Schmidt, R.; Fried, L. *Munitions Technology Development Program*; UCRL-ID-103483-99; 1999; p. II-5.
6. Wang, Y.; Liu, Y.; Song, S.; Yang, Z.; Qi, X.; Wang, K.; Liu, Y.; Zhang, Q.; Tian, Y. Accelerating the Discovery of Insensitive High-Energy-Density Materials by a Materials Genome Approach. *Nat. Commun.* **2018**, *9*, 2444. [CrossRef]
7. Li, C.; Lei, C.; Tang, J.; Zhu, T.; Cheng, G.; Yang, H. C–C bonded bis-5,6 fused triazole–triazine compound: An advanced heat-resistant explosive with high energy and low sensitivity. *Dalton Trans.* **2022**, *51*, 15292–15299. [CrossRef]
8. Zheng, X.; Xue, Y.; Dai, C.; Yang, H.; Cheng, G. The skeleton of 5,7-fused bicyclic imidazole-diazepine for heat-resistant energetic materials. *Def. Technol.* **2023**, *27*, 193–199. [CrossRef]
9. Tang, Y.; He, C.; Imler, G.H.; Parrish, D.A.; Shreeve, J.M. Aminonitro Groups Surrounding a Fused Pyrazolotriazine Ring: A Superior Thermally Stable and Insensitive Energetic Material. *ACS Appl. Energy Mater.* **2019**, *2*, 2263–2267. [CrossRef]
10. Li, J.; Liu, Y.; Ma, W.; Fei, T.; He, C.; Pang, S. Tri-Explosophoric Groups Driven Fused Energetic Heterocycles Featuring Superior Energetic and Safety Performances Outperforms HMX. *Nat. Commun.* **2022**, *13*, 5697. [CrossRef]
11. Reed, A.E.; Weinstock, R.B.; Weinhold, F. Natural population analysis. *J. Chem. Phys.* **1985**, *83*, 735–746. [CrossRef]
12. Lu, T.; Chen, F. Bond Order Analysis Based on the Laplacian of Electron Density in Fuzzy Overlap Space. *J. Phys. Chem. A* **2013**, *117*, 3100–3108. [CrossRef] [PubMed]

13. Lu, T.; Chen, Q.; Liu, Z. A Thorough Theoretical Exploration of Intriguing Characteristics of Cyclo [18]carbon: Geometry, Bonding Nature, Aromaticity, Weak Interaction, Reactivity, Excited States, Vibrations, Molecular Dynamics and Various Molecular Properties. *ChemRxiv* **2019**. [CrossRef]
14. Schmider, H.L.; Becke, A.D. Chemical content of the kinetic energy density. *J. Mol. Struct. THEOCHEM* **2000**, *527*, 51–61. [CrossRef]
15. Lu, T.; Chen, Q. A simple method of identifying π orbitals for non-planar systems and a protocol of studying π electronic structure. *Theor. Chem. Acc.* **2020**, *139*, 25. [CrossRef]
16. Giambiagi, M.; De Giambiagi, M.S.; Mundim, K.C. Definition of a multicenter bond index. *Struct. Chem.* **1990**, *1*, 423–427. [CrossRef]
17. Weininger, D. SMILES, a Chemical Language and Information System. 1. Introduction to Methodology and Encoding Rules. *J. Chem. Inf. Model.* **1988**, *28*, 31–36. [CrossRef]
18. O'Boyle, N.M.; Banck, M.; James, C.A.; Morley, C.; Vandermeersch, T.; Hutchison, G.R. Open Babel: An Open Chemical Toolbox. *J. Cheminform.* **2011**, *3*, 33. [CrossRef]
19. Gaussian 09 Revision D.01; Frisch, M.J.; Trucks, G.W.; Schlegel, H.B.; Scuseria, G.E.; Robb, M.A.; Cheeseman, J.R.; Scalmani, G.; Barone, V.; Petersson, G.A.; et al. *Gaussian 09 Revision D.01*; Gaussian Inc.: Wallingford, CT, USA, 2016.
20. Lu, T.; Chen, F. Multiwfn: A Multifunctional Wavefunction Analyzer. *J. Comput. Chem.* **2012**, *33*, 580–592. [CrossRef] [PubMed]
21. Lu, T.; Chen, F. Quantitative analysis of molecular surface based on improved Marching Tetrahedra algorithm. *J. Mol. Graph. Model.* **2012**, *38*, 314–323. [CrossRef]
22. Neese, F. The ORCA Program System. *WIREs Comput Mol. Sci.* **2012**, *2*, 73–78. [CrossRef]

molecules

Review

Advances in Synthesis and Ignition Performance of Ionic Liquid–Hydrogen Peroxide Green Propellants

Yongting Zhang [1,2,†], Xing Zhang [1,2,†], Dangyue Yin [1,2,*] and Qinghua Zhang [1,2]

1 School of Astronautics, Northwestern Polytechnical University, Xi'an 710065, China; 18091925902@163.com (Y.Z.); zhangxing@nwpu.edu.cn (X.Z.); qinghuazhang@nwpu.edu.cn (Q.Z.)
2 National Key Laboratory of Solid Propulsion, Xi'an 710065, China
* Correspondence: yindangyue@nwpu.edu.cn; Tel.: +86-029-88495827
† These authors contributed equally to this work.

Abstract: The ionic liquid–hydrogen peroxide propellant system has emerged as a promising green propellant candidate, synergistically combining the unique advantages of ionic liquids (such as negligible vapor pressure, low melting points, high thermal stability and structural tunability) with the merits of hydrogen peroxide (including high density, low volatility, minimal viscosity, reduced corrosivity, and environmentally benign decomposition products). In this work, we provide a comprehensive review of the synthesis strategies and ignition performance of the ionic liquid–hydrogen peroxide propellant system, systematically categorizing them into two classes: "self-igniting propellants" and "promoter-dependent propellants". This review emphasizes the critical role of anion-specific design and catalytic engineering in advancing the performance of ionic liquid–hydrogen peroxide propellant systems, while also addressing the current challenges and future directions in this rapidly evolving field.

Keywords: liquid propellant; ionic liquids; hydrogen peroxide; ignition delay

1. Introduction

Propellants, defined as energetic materials that undergo rapid combustion or decomposition to generate high-pressure gases for thrust production, play a pivotal role in aerospace propulsion, satellite trajectory control, and gas generation systems [1]. While solid propellants dominate certain applications, liquid propellants offer superior thrust modulation, enhanced combustion reliability, and restart capability, making them indispensable for advanced rocket engines and spacecraft [2,3]. Liquid propellants can be categorized into two types: monopropellants and bipropellants. Unlike monopropellants, bipropellants do not require an external ignition device, which simplifies the engine structure, reduces costs, enhances combustion reliability, and allows for multiple restarts and shutdowns to improve flexibility [4].

The bipropellant consists of two components: an oxidizer and a fuel. The oxidizers are usually strong oxidizing compounds such as nitric acid, dinitrogen tetroxide, white fuming nitric acid, and red nitric acid [5]. These substances are not only highly corrosive and volatile but also produce significant amounts of nitrogen oxides upon combustion [6], posing severe toxicity risks to human health and causing substantial environmental pollution. Hydrazine and its derivatives exhibit excellent performance as propellant fuel and have widespread applications in liquid propellant systems, but they are explosive and highly toxic chemicals with a low boiling point and strong volatility. The use of hydrazine compounds will cause many destructive consequences to human health and ecological

https://doi.org/10.3390/molecules30081789

environments, and in practical applications, the transportation cost of such substances is high, and the storage is difficult [7–10].

This critical safety–environmental paradox has driven intensive research toward green liquid propellants. Ionic liquids, a class of salt compounds composed of organic cations and organic or inorganic anions usually with a melting point below 100 °C, stand out for their low toxicity, non-volatility, short ignition delay, structural tunability, and high safety [11,12]. Hydrogen peroxide (H_2O_2) is a common industrial chemical and environmentally friendly oxidizer which is produced in extremely large quantities. It has many advantages of low toxicity, low vapor pressure, and non-toxic decomposition products [13,14]. In addition, there have been significant improvements in its stability and storability [15]. Therefore, liquid propellants based on the combination of ionic liquid–hydrogen peroxide are expected to realize the true greening of self-ignition liquid propellants. Therefore, developing such propellants is crucial for promoting the green process of space power in the future.

In fact, only a limited number of existing energetic ionic liquids, primarily those incorporating borohydride or thiocyanate anions, exhibit spontaneous hypergolic ignition with high concentration of H_2O_2. For ionic liquids that do not exhibit this capability, catalytic promoters, such as iodine or copper-based organometallics, have been strategically employed to achieve millisecond-scale ignition delays [16]. This review systematically categorizes ionic liquid–hydrogen peroxide propellant systems into two distinct classes, "self-igniting propellants" and "promoter-dependent propellants", with a focus on four anionic families: borohydrides, thiocyanates, dicyanamide, and other emerging anions. We critically summarized their synthesis pathways, structure–ignition performance relationships, and catalytic effects, aiming to establish development ideas and explore application directions for next-generation green propellants.

2. Ionic Liquid–Hydrogen Peroxide Liquid Propellant Capable of Self-Ignition

2.1. ILs Based on Borohydride-Rich Anions

Among the existing ionic liquids, only a small amount of ionic liquids can directly and rapidly self-ignite with a high concentration of H_2O_2, which usually have borohydride bonds with strong reducing ability.

In 2011, Schneider et al. first proposed studying the spontaneous combustion behavior of ionic liquids in H_2O_2 [17]. They speculated that ionic liquids with metal hydride anions could show high solubility of H_2O_2 in ether and prepared ionic liquid borohydride with subsequent conversion to $[Al(BH_4)_4]^-$ ionic liquids. Trihexyltetradecylphosphonium (THTDP) chloride was utilized to enable facile, quantitative anion exchange of Cl^- for BH_4^-. The resulting new material $[THTDP][BH_4]$ (IL-1) was a viscous ionic liquid. Subsequently, another ionic liquid $[THTDP][AI(BH_4)_4]$ (IL-2) was synthesized by reacting with a slight excess of aluminum borohydride. The structures of IL-1 and IL-2 are shown in Figure 1. The two ionic liquids were then ignited with 90% and 98% H_2O_2, respectively, demonstrating their ability to undergo spontaneous combustion with H_2O_2 without a catalytic promoter. Remarkably, IL-2 exhibits fast self-ignition performance with both 90% and 98% H_2O_2, with ignition delay times (IDTs) less than 30 ms, while the IDT of IL-1 was longer than 3 s. These tests confirm that ILs based on borohydride-rich anions are universally reactive, offering a new approach for the designing of novel self-ignition ionic liquids with H_2O_2. However, the harsh synthesis conditions and poor water stability of these ionic liquids seriously restrict their extensive application as propellant fuels.

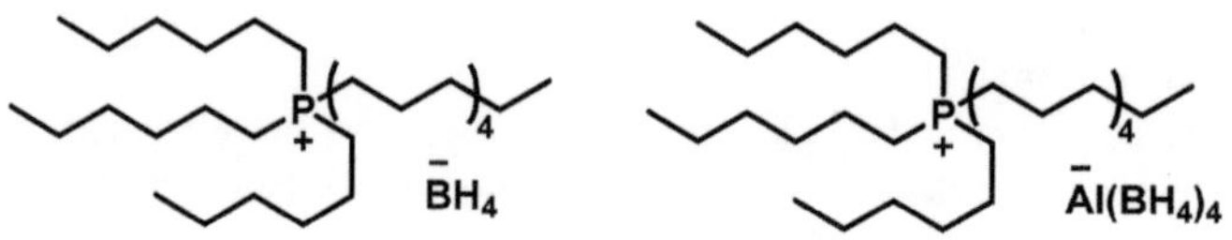

IL−1 IL−2

Figure 1. Structures of [THTDP][BH_4] (IL-1) and [THTDP][$Al(BH_4)_4$] (IL-2) ionic liquids.

In 2017, Bhosale et al. conducted a theoretical investigation on the performance parameters of 14 ionic liquids (IL-3 to IL-16, Figure 2) with different oxidizers including H_2O_2, using the NASA-CEC-71 program [18]. As shown in Table 1, the effect of their performance parameters was analyzed in detail and compared with 1,1-dimethylhydrazine (UDMH). The results indicated that all ionic liquids exhibited superior performance with H_2O_2 compared to other oxidizers. Specifically, the ionic liquids containing borohydride-rich anions demonstrated higher density-specific impulse (ρI_{sp}) values than UDMH. Among them, IL-3 exhibited the highest ρI_{sp} out of all the oxidizers as it had higher density than any other ionic liquid, achieving a ρI_{sp} of 319.8 g s cm^{-3} with H_2O_2. Therefore, combinations of ionic liquids based on borohydride-rich anions and H_2O_2 oxidizers can be recommended as future green propellants. However, since these experimental data were based on theoretical calculations and no actual drop tests were conducted, further verification of the spontaneity of these ionic liquids is necessary. Additionally, unless specifically noted, all IDT data presented in this study were obtained through experimental measurements.

In order to further study the above research achievements, Bhosale et al. synthesized several green hypergolic fuels with borohydride bond-rich anions, and investigated their hypergolic reactivity with 95% H_2O_2 [19]. As shown in Figure 2, the ionic liquids included [EMIM][BH_3CN] (IL-3), [AEIM][BH_3CN] (IL-4) and [EMIM][BH_4] (IL-17), which were synthesized by simple metathesis reactions involving halide salts and metal borohydrides in selective solvents (CH_2Cl_2 or MeCN) at room temperature. IL-17 exhibited a short IDT of 18.5 ms with 95% H_2O_2, while IL-3 and IL-4 exhibited longer IDTs (>1000 ms) with 95% H_2O_2 in both fuel-rich and oxidizer-rich conditions. However, IL-3 and IL-4 were liquid, but IL-17 was solid at room temperature. Furthermore, the theoretical specific impulse (I_{sp}) of ionic liquids was calculated with 95% H_2O_2 using NASA CEA 400 software under specified conditions, considering a range of oxidizer-to-fuel ratios (O/F, 0.5–5.5). IL-3 and IL-4 showed the lowest I_{sp} of 253 s at an O/F of 3.5 and had a fairly low viscosity (η of 17 and 19 mPa s), good thermal stability (T_d of 265 and 247 °C), and acceptable density (ρ of 0.95 and 0.98 g cm^{-3}), suggesting that ILs based on borohydride-rich anions have application prospects as H_2O_2-based green spontaneous combustion propellants.

In 2023, Wang et al. synthesized four novel hypergolic fluids based on borohydride ionic liquids [1-ethyl-3-methylimidazolium borohydride ([EMIM][BH_4]) or 1-butyl-3-methylimidazolium borohydride ([BMIM][BH_4])] by an in situ synthetic method in an organic super-base [1,5-Diazabicyclo[4.3.0]-5-nonene (DBN) or 1,8-Diazabicyclo[5.4.0]undec-7-ene (DBU)], as shown in Figure 3 [20]. All these hypergolic fuels exhibited high densities and low viscosities due to the incorporation of super-bases with high densities and low viscosities, which show significantly superior properties compared to those of pure [BMIM][BH_4] (ρ of 0.91 g cm^{-3} and η of 486.6 mPa s). Additionally, the hypergolic fluids with 90% H_2O_2 demonstrated an acceptable IDT, with a minimum of 28.3 ms, which is lower than that with WFNA. That is attributed to the formation of a homogeneous mixed layer and no occurrence of secondary rebounds during drop tests conducted with H_2O_2, indicating their strong potential as green fuels for use in green propellant systems. The organic super-base acted as an important solvent for the simple and low-cost synthesis of borohydride-containing ionic liquids; meanwhile, it is also the main component of fuels that possess high density and low viscosity, and the ionic liquids in DBN and DBU acted as triggers to contribute to the self-ignition of DBN and DBU with 90% H_2O_2.

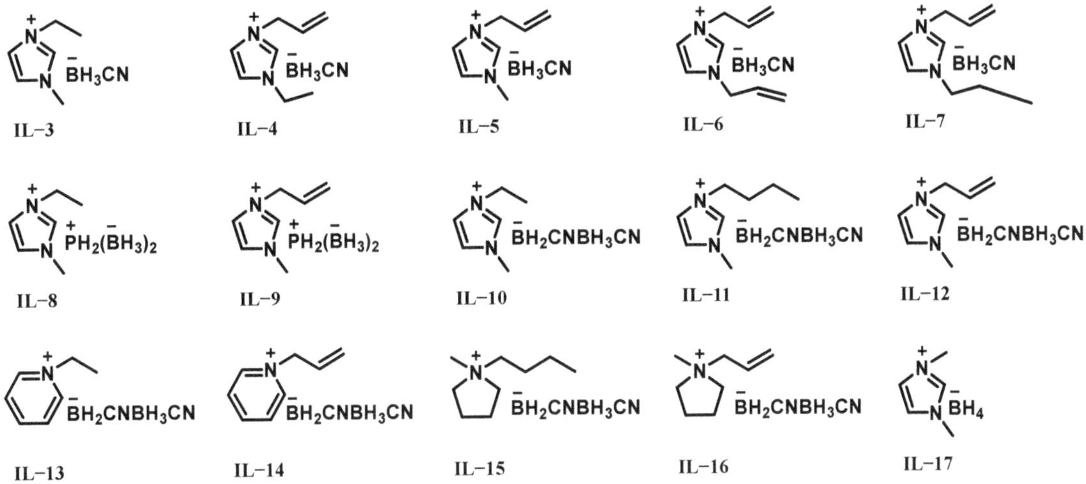

Figure 2. Structures of IL-3 to IL-17.

Table 1. Theoretical performances of 14 ionic liquids with H_2O_2.

Hypergolic Fuels	O/F [a]	Vc [b] (m s^{-1})	I_{sp} [c] (s)	I_{vac} [d] (s)	ρ [e] (g cm^{-3})	ρI_{sp} [f] (s g cm^{-3})
IL-3	3.5	1670	244.1	273.3	1.310	319.8
IL-4	3.5	1663	243.1	272.1	1.299	315.9
IL-5	3.5	1673	244.5	273.7	1.303	318.6
IL-6	3.5	1668	243.9	272.9	1.301	317.4
IL-7	3.5	1669	244.0	272.9	1.294	315.8
IL-8	2.5	1664	243.2	271.9	1.234	300.2
IL-9	2.5	1666	243.5	272.2	1.240	301.9
IL-10	3.5	1646	240.6	269.6	1.288	309.9
IL-11	3.5	1648	240.9	269.7	1.297	312.5
IL-12	3.5	1650	241.1	270.2	1.292	311.4
IL-13	3.5	1646	240.6	269.4	1.291	310.7
IL-14	3.5	1642	240.0	268.8	1.294	310.7
IL-15	4.0	1659	242.4	271.5	1.286	311.8
IL-16	4.0	1661	242.7	272.0	1.292	313.7
UDMH	3.5	1694	247.7	276.9	1.224	303.3

[a] optimum oxidizer-to-fuel ratio of propellant. [b] characteristic velocity of propellant. [c] specific impulse of propellant. [d] vacuum-specific impulse. [e] density of propellant. [f] density-specific impulse of propellant. (These parameters were determined under frozen flow conditions by assuming chamber pressure (P_c, 2.4 × 10^6 Pa), exit pressure (P_e, 9.8 × 10^4 Pa), and the area ratio (A_e/A_t, 4)).

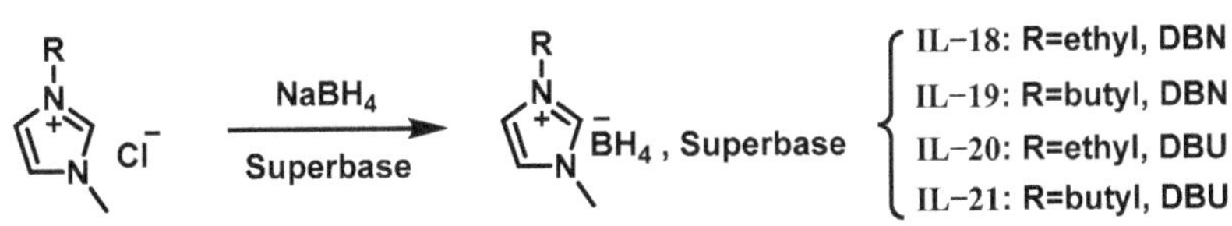

Figure 3. Synthesis method of the composite hypergolic fuels.

The physicochemical properties of all ionic liquids based on borohydride-rich anions are shown in Table 2. To date, the anions of borohydride-based ionic liquids that can self-ignite mainly include $[BH_4]^-$, $[BH_3CN]^-$ and a singular example of $[Al(BH_4)_4]^-$. Although theoretical calculations have demonstrated that the ionic liquids containing $[BH_2CNBH_3CN]^-$ and $[PH_2(BH_3)_2]^-$ also exhibit high performances with H_2O_2 and have higher ρI_{sp} values than UDMH, they were not confirmed by drop-test experiments. Among all these ILs, IL-17 exhibited the shortest IDT of 18.5 ms and a competitive I_{sp} of 258 s, but it was in a solid state at ambient temperatures. IL-18 to IL-21 had high densities and self-ignition behavior owing to the solvents (DBN and DBU), which were the main components of the fuels. These limitations highlight two research priorities for borohydride-rich anion architectures: experimentally validating computationally predicted hypergolic candidates, and developing novel ionic liquids that combine room-temperature liquidity, solvent-free formulations, and spontaneous ignition capabilities.

Table 2. Physicochemical properties of all ionic liquids based on borohydride-rich anions.

IL	ρ [a] (g cm^{-3})	η [b] (mPa s)	T_d [c] (°C)	T_m [d] (°C)	ΔH_f [e] (kJ mol^{-1})	I_{sp} [f] (s)	O/F [g]	IDT [h] (ms)	State [i]	Oxidizer
IL-1	-	-	-	-	-	-	-	>3000	liquid	98% H_2O_2, 90% H_2O_2
IL-2	-	-	-	-	-	-	-	<30	liquid	98% H_2O_2, 90% H_2O_2
IL-3	0.98	19	247	−71	136	253	3.5	>1000	liquid	95% H_2O_2
IL-4	0.95	17	265	<−50	225	253	3.5	>1000	liquid	95% H_2O_2
IL-17	0.92	-	130	50	132	258	3.5	18.5	soild	95% H_2O_2
IL-18	1.02	58	-	-	-	-	-	28.3	liquid	90% H_2O_2
IL-19	1.03	34	-	-	-	-	-	86.8	liquid	90% H_2O_2
IL-20	1.01	110	-	-	-	-	-	344	liquid	90% H_2O_2
IL-21	1.00	81	-	-	-	-	-	127	liquid	90% H_2O_2

[a] density at 25 °C. [b] viscosity at 25 °C. [c] decomposition temperature. [d] melting temperature. [e] heat of formation. [f] specific impulse (under frozen flow conditions, $P_c = 2.4 \times 10^6$ Pa, $P_e = 9.8 \times 10^4$ Pa, $A_e/A_t = 4$). [g] oxidizer-to-fuel ratio. [h] ignition delay time. [i] physical state of the propellant at room temperature.

2.2. ILs Based on Thiocyanate Anions

Recently, researchers have discovered that ionic liquids based on thiocyanate anions can also self-ignite with H_2O_2. In 2021, Lauck et al. conducted drop tests with two different ionic liquids based on thiocyanate anions [1-ethyl-3-methylimidazolium thiocyanate ([EMIM][SCN], IL-22) and 1-butyl-3-methylimidazolium thiocyanate ([BMIM][SCN], IL-23)] and H_2O_2 [21]. The structures of IL-22 and IL-23 ionic liquids are shown in Figure 4. Both ionic liquids can be ignited with 96.1% H_2O_2, exhibiting average IDTs of 31.7 ms for IL-22 and 45 ms for IL-23. Additionally, the theoretical I_{sp} calculated with NASA CEA for the two ionic liquids was approximately 318 s. Owing to their higher fuel density, the ρI_{sp} was enhanced by 10% compared to the conventional hypergolic propellant combination of monomethyl hydrazine/dinitrogen tetroxide (MMH/NTO). These results demonstrate that ionic liquids based on thiocyanate anions combined with high-concentration H_2O_2 hold great potential as viable alternatives to existing hypergolic propellant systems.

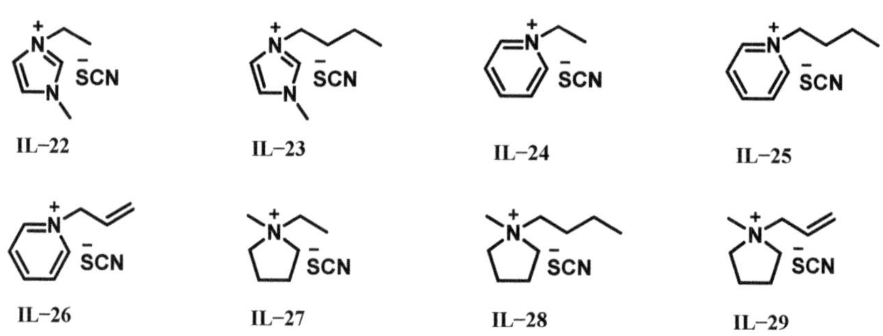

Figure 4. Structures of IL-22 to IL-29.

In 2021, Ricker et al. proposed six pyridinium- and pyrrolidinium-based thiocyanate ionic liquids (IL-24 to IL-29) as new fuel candidates with 97.4% H_2O_2 as an oxidizer, as shown in Figure 4 [22]. The results indicated that all ionic liquids with pyridinium-based cations showed shorter IDTs than those with cationic pyrrolidinium-based frameworks. Within the groups of the same heterocycles, shorter alkyl chains in the cation positively influenced the ignition behavior of the hypergolic propellants, while the degree of unsaturation in the side chains had no significant effect on IDTs. Furthermore, all studied fuels exhibited higher theoretical ρI_{sp} compared to the conventional hypergolic propellant combination of MMH/NTO. IL-24 demonstrated the shortest IDT (26.8 ms), the lowest viscosity (27.5 mPa s), a high density (1.13 g cm^{-3}), and excellent thermal stability (T_d of 247 °C). This study not only identifies IL-24 as a promising candidate for future orbital propulsion systems but also provides new insights into the role of cationic structures in ionic liquids for hypergolic propellant combinations with H_2O_2.

The following year, Ricker et al. synthesized seven protic ionic liquids (IL-30 to IL-36) containing thiocyanate anions, as illustrated in Figure 5 [23]. Theoretical calculations using the NASA CEA code revealed that all ionic liquids with 97% H_2O_2 exhibited higher ρI_{sp} compared to the conventional toxic propellant MMH/NTO. The results showed that the IDTs increased with increasing chain length at the same position, which can be attributed to steric effects. Among these, IL-30 demonstrated an exceptionally short average IDT of 7.3 ms. However, since IL-30 is solid at ambient conditions, a blend of 35 wt% IL-30 and 65 wt% IL-22 (liquid at room temperature) was developed, forming a novel liquid bipropellant named HIM_35. This mixture achieved a low IDT of 16.7 ms. Notably, these fuels avoid the use of catalytic transition metals, boron-based compounds (which generate solid combustion residues), or air/moisture-sensitive hydrides. Consequently, they represent highly promising alternatives for environmentally friendly hypergolic bipropellants in space propulsion systems.

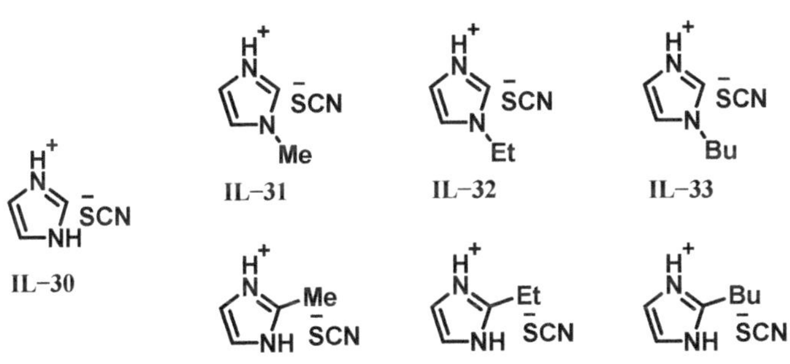

Figure 5. Structures of IL-30 to IL-36.

In 2024, Stölzle et al. synthesized three trialkylsulfonium thiocyanate ionic liquids (IL-37 to IL-39, Figure 6) and evaluated their potential as hypergolic fuels with high-concentration H_2O_2 [24]. Theoretical combustion calculations revealed that IL-39 achieved the highest I_{sp} of 314 s at O/F of 4.5, while IL-37 exhibited superior ρI_{sp} of 429 s g cm^{-3} at O/F = 3.9, attributed to its exceptional density. Hypergolic ignition behavior was further investigated through drop tests under ambient conditions and at reduced fuel temperatures (1 °C and −25 °C). All three propellant combinations maintained reliable ignition even at −25 °C, though lower temperatures prolonged IDTs. Notably, IL-38 demonstrated the shortest IDT, with average values of 30.8 ms at ambient conditions, increasing marginally to 32.8 ms at 1 °C and significantly to 51.8 ms at −25 °C, highlighting temperature-dependent kinetic limitations in thiocyanate-based systems.

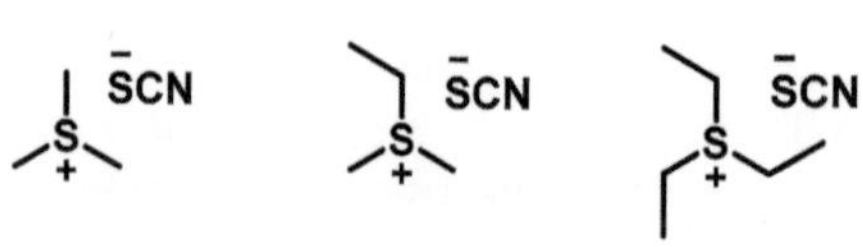

IL−37 **IL−38** **IL−39**

Figure 6. Structures of IL-37 to IL-39.

The physicochemical properties of all ionic liquids based on thiocyanate anions are shown in Table 3. These ionic liquids exhibit rapid hypergolic ignition with high-concentration H_2O_2, achieving a minimum IDT of 7.3 ms and a high I_{sp} of up to 320 s. Cationic structural design plays a critical role in their performance. Shorter alkyl chains on pyridinium, pyrrolidinium, and imidazolium cations significantly reduce IDTs. However, these systems are heavily dependent on the high-concentration H_2O_2, and lower temperatures markedly degrade the ignition efficiency (e.g., IL-37's IDT increases to 83.6 ms at −25 °C). Notably, trialkylsulfonium-based IL-38 defies the alkyl chain length trend, achieving the shortest IDT of 30.8 ms despite longer substituents, suggesting potential asymmetry effects in cationic frameworks.

Table 3. Physicochemical properties of all ionic liquids based on thiocyanate anions.

IL	ρ [a] (g cm^{-3})	η [b] (mPa s)	T_d [c] (°C)	T_m [d] (°C)	ΔH_f [e] (kJ mol^{-1})	I_{sp} [f] (s)	O/F [g]	IDT [h] (ms)	State [i]	Oxidizer
IL-22	1.11	23	-	−6	53	317	3.8	31.7	liquid	96.1% H_2O_2
IL-23	1.07	36	-	−29	−5	319	4.2	45	liquid	96.1% H_2O_2
IL-24	1.13	28	247	-	−149	313	4.3	26.8	liquid	97.4% H_2O_2
IL-25	1.09	87	258	-	−126	317	4.5	33.9	liquid	97.4% H_2O_2
IL-26	1.14	38	197	-	93	317	4.2	29.5	liquid	97.4% H_2O_2
IL-27	1.07	-	264	55	−162	320	4.5	43.1	solid	97.4% H_2O_2
IL-28	1.05	549	261	-	−477	318	4.9	61.9	liquid	97.4% H_2O_2
IL-29	1.07	81	232	-	−140	319	4.6	48.9	liquid	97.4% H_2O_2
IL-30	1.36	-	219	106	127	309	3.0	7.3	solid	97% H_2O_2
IL-31	1.27	-	229	45	58	312	3.4	23.0	solid	97% H_2O_2
IL-32	1.14	77	237	2	49	314	3.6	42.8	liquid	97% H_2O_2
IL-33	1.09	170	250	−18	11	318	4.0	46.1	liquid	97% H_2O_2
IL-34	1.28	-	252	84	−14	310	3.4	16.5	solid	97% H_2O_2
IL-35	1.24	-	259	87	−73	312	3.7	20.2	solid	97% H_2O_2
IL-36	1.08	126	270	−74	−2	318	4.0	28.4	liquid	97% H_2O_2
IL-37	1.26	-	116	66	−176	307	3.0	55.1 (21 °C); 65.1 (1 °C); 83.6 (−25 °C)	solid	97% H_2O_2
IL-38	1.13	26	119	0	−170	311	4.1	30.8 (21 °C); 32.8 (1 °C); 51.8 (−25 °C)	liquid	97% H_2O_2
IL-39	1.09	34	123	−13	−269	314	4.5	61.8 (21 °C); 69.2 (1 °C); 79.2 (−25 °C)	liquid	97% H_2O_2

[a] density at 25 °C. [b] viscosity at 25 °C. [c] decomposition temperature. [d] melting temperature. [e] heat of formation. [f] specific impulse (under frozen flow conditions, P_c = 1 MPa, A_e/A_t = 330). [g] oxidizer-to-fuel ratio. [h] ignition delay time. [i] physical state of the propellant at room temperature.

Future investigations should prioritize these objectives: the first is elucidating the structure–ignition relationship in trialkylsulfonium cations to clarify the role of asymmetry, and the second is expanding the library of hypergolic thiocyanate-based ionic liquids through systematic cation–anion pairing strategies. Additionally, current systems require high-concentration H_2O_2 for spontaneous ignition, so further research is essential to evaluate their compatibility with low-concentration H_2O_2, which could enhance practicality and environmental adaptability.

3. Ionic Liquid–Hydrogen Peroxide Liquid Propellant Requiring Promoters

3.1. Promoters for ILs Based on Borohydride-Rich Anions and Hydrogen Peroxide

In 2018, Chinnam et al. developed novel iodine-rich hypergolic promoters (ILP-1 and ILP-2) to enable rapid ignition between IL-3 and H_2O_2 [25]. As illustrated in Figure 7, these promoters were synthesized to address the extremely long IDT (>30 s) observed for IL-3 with 70% H_2O_2 in the absence of a promoter. Remarkably, adding 8 wt% ILP-2 reduced the IDT to 45 ms with 70% H_2O_2 and further to 17 ms with 95% H_2O_2. In order to better understand the promoter effect of the $[B_{12}I_{12}]^{2-}$ anion on this oxidation process, theoretical calculations were conducted, and they revealed that the $[B_{12}I_{12}]^{2-}$ anion in ILP-2 significantly lowered the Gibbs free energies of intermediates during the initial oxidation steps, thereby accelerating substrate decomposition. Thus, organometallic salts containing ferrocene ($[FcCH_2NEtMe_2]^+$) or copper ($[Cu(en)_2(CH_3CN)_2]^{2+}$) exhibited enhanced catalytic activity when paired with $[B_{12}I_{12}]^{2-}$, forming unique bifunctional promoters. This multifunctional design demonstrates a breakthrough in utilizing safer 70% H_2O_2 (instead of highly concentrated H_2O_2) for hypergolic systems, offering a sustainable pathway for developing green oxidizers in space propulsion.

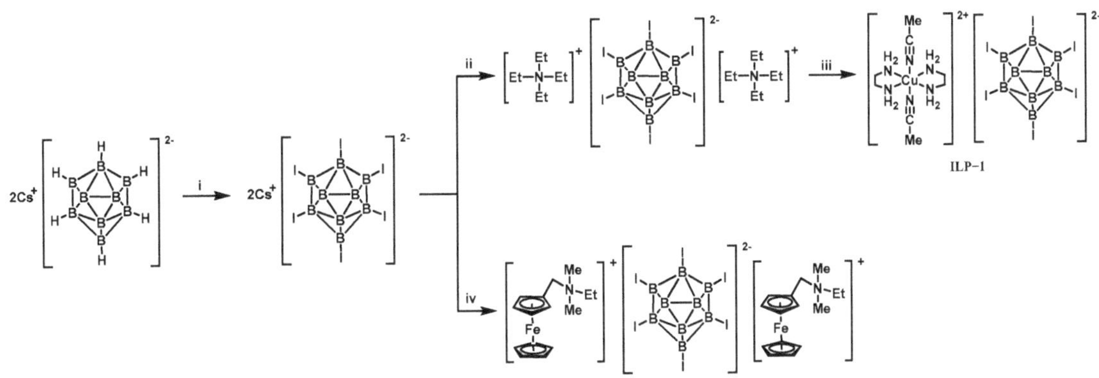

Figure 7. Synthesis scheme of ILP-1 and ILP-2. (Experimental conditions: (i) I_2; (ii) tetraethylammonium bromide; (iii) $[Cu(en)_2(OH)_2]$; (iv) $[FeCH_2NEtMe][I]$).

In the same year, Wang et al. synthesized and characterized four iodocuprate-based ionic liquid promoters (ILP-3 to ILP-6) to achieve rapid ignition between fuels (IL-3 or $[MIM][BH_3]$ (IL-40)) and 95% H_2O_2 [26]. The synthesis routes of these promoters are shown in Figure 8. Among them, ILP-5 emerged as the most promising candidate. Firstly, its decomposition temperature exceeded that of the fuels, ensuring stability in fuel–promoter mixtures for weeks. Secondly, it maintained the mixture's viscosity at 50 mPa s and remained homogeneously dispersed in IL-3 for over four weeks without degradation of either the promoter or fuel. Furthermore, with 10 wt% ILP-5, the IDTs of IL-3 and IL-40 were shortened to 24 ms and 14 ms, respectively. Regarding the reaction mechanism of the catalytic process, the promoters can react with H_2O_2, accelerating substrate decomposition. In summary, the exceptional stability and catalytic efficiency of the newly synthesized promoters, especially of ILP-5, highlight their potential to advance the development of green bipropellant systems using H_2O_2 for space propulsion applications.

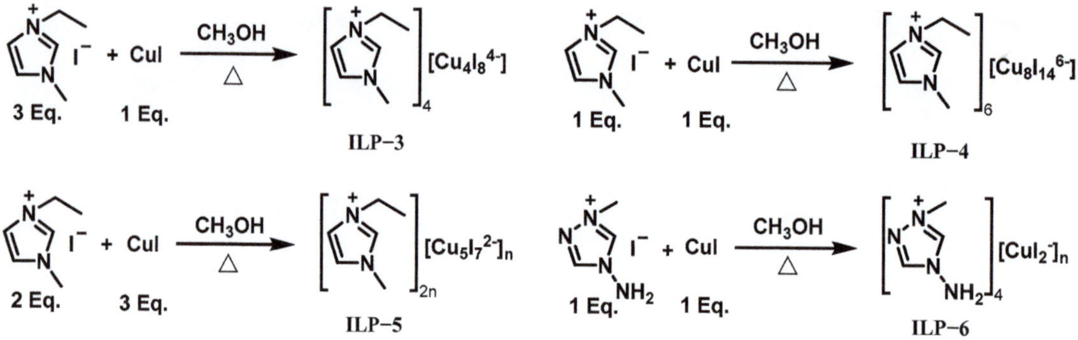

Figure 8. Synthesis routes of ILP-3 to ILP-6.

In 2019, Wang et al. synthesized a series of hypergolic ionic liquids (IL-41 to IL-49) containing the cyano (1H-1,2,3-triazole-1-yl) dihydroborate anion [27]. The synthesis routes of these ionic liquids are shown in Figure 9. While the pure ionic liquids did not spontaneously ignite with 90% H_2O_2, the addition of 15 wt% iodine (I_2, ILP-7) enabled the hypergolic ignition of IL-42 to IL-43. This result demonstrated that iodine acted as an efficient catalyst to promote reactions between boron-containing ionic liquids and H_2O_2. Such catalytic capability positions iodine as a promising candidate for green propellant systems based on ionic liquid–H_2O_2 combinations.

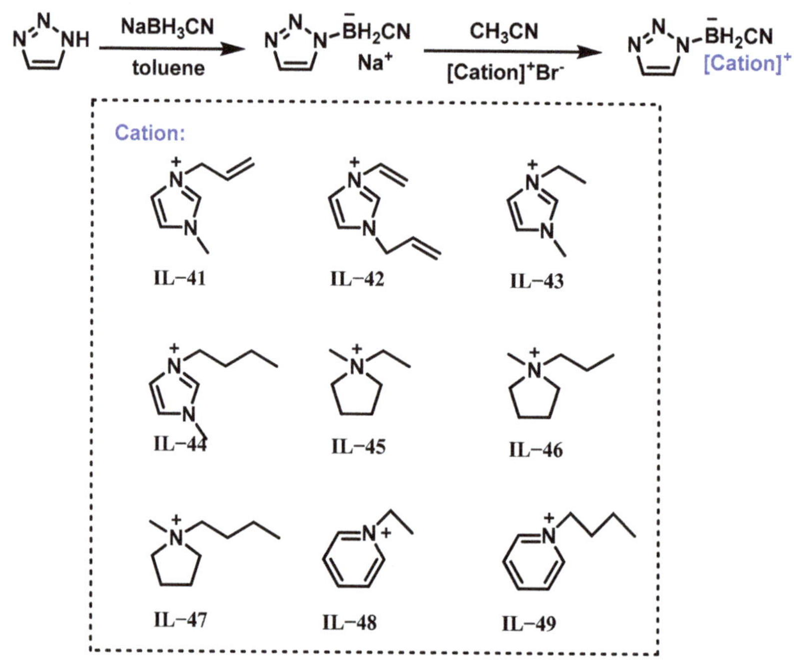

Figure 9. Synthesis of IL-41 to IL-49.

In 2020, Bhosale et al. investigated the use of various promoters (ILP-8 to ILP-12) to enhance the hypergolic ignition of ionic liquids (IL-3 and IL-4) with 95% H_2O_2 [5]. The structures of ILP-8 to ILP-11 are shown in Figure 10. At a concentration of 5 wt%, these promoters reduced the IDTs of IL-3 to 139 ms (ILP-8), 395 ms (ILP-9), 887 ms (ILP-10), and 73 ms (ILP-11). 1,3-dimethyl imidazolium copper iodide ($[diMIM]_n[Cu_2I_3]_n$, ILP-12) was a newly synthesized promoter, shown in Figure 11, and its physicochemical properties were further enhanced. Adding 5 wt% ILP-12 shortened the IDT of IL-3 to 87 ms. Notably, increasing the ILP-12 concentration from 2 to 15 wt% in IL-3 significantly reduced the IDT of IL-3 from 126 to 29 ms and 47 to 13 ms under oxidizer-rich and fuel-rich conditions, respectively. Meanwhile, higher ILP-12 concentrations effectively improved the density and viscosity of the fuel mixture. These findings demonstrate that additive-promoted hypergolic combustion offers a viable pathway to replace conventional toxic propellant systems.

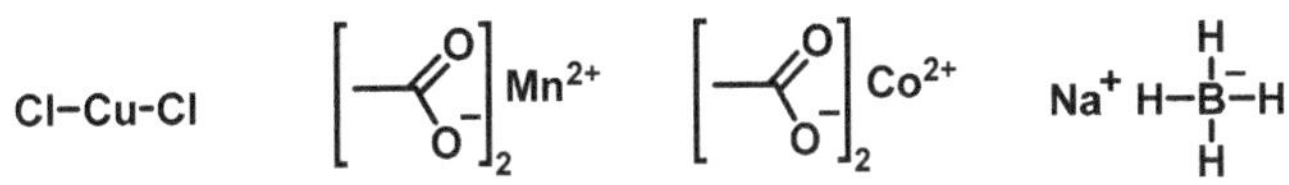

ILP−8 ILP−9 ILP−10 ILP−11

Figure 10. Structures of ILP-8 to ILP-11.

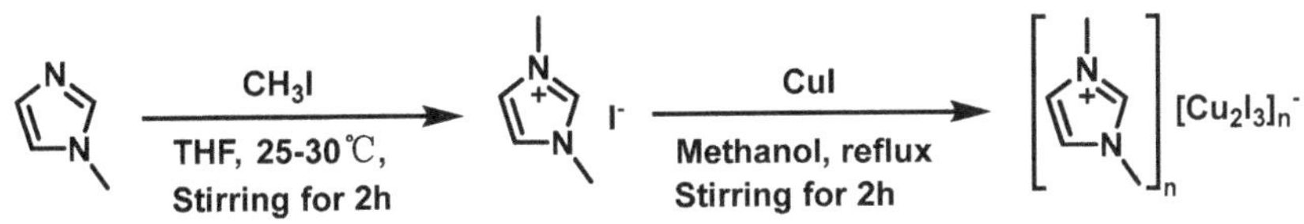

ILP−12

Figure 11. Synthesis of ILP-12.

In 2020, addressing the poor solubility of existing hypergolic promoters, Wang et al. synthesized four novel promoters (ILP-13 to ILP-16) with ultrahigh solubility in energetic ionic liquids based on the principle of similarity compatibility [28]. Their structures are shown in Figure 12. To systematically compare the roles of anions and cations, two reference promoters were designed: The first is the "cation active" copper-free reference promoter ILP-15, whereas the second is an "anion active" iron-free reference promoter ILP-16. The ignition effect of these promoters on ionic liquid fuels IL-3 and IL-5 was studied. Adding 10 wt% ILP-13 or ILP-14 reduced the IDT of IL-3 to 38 ms and 31 ms, respectively, while for IL-5, the IDTs were 89 ms and 56 ms under the same conditions. In addition, ILP-13 and ILP-14 exhibited excellent solubility and chemical stability in ionic liquids. This study highlights that incorporating metals (e.g., Fe, Cu) into the anion of the promoter plays a predominant role in reducing IDTs, providing critical insights for designing next-generation hypergolic systems.

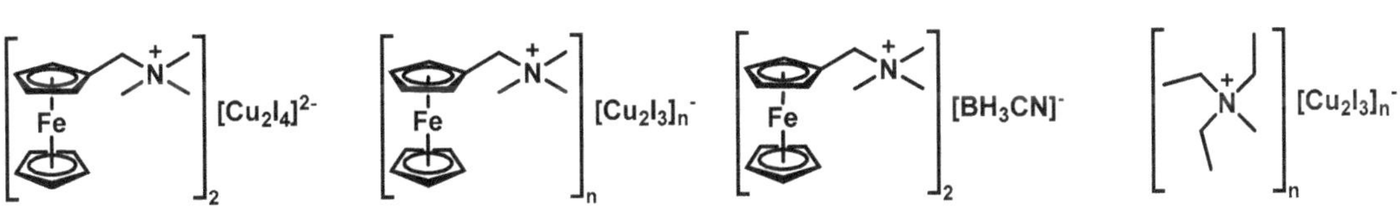

ILP−13 ILP−14 ILP−15 ILP−16

Figure 12. Structures of ILP-13 to ILP-16.

In the same year, Bhosale et al. conducted research on the use of sodium iodide (NaI, ILP-17) as a promoter to enhance the ignition performance of the hypergolic fuels IL-3 and IL-40, with 95% H_2O_2 serving as the green oxidizer [29]. They prepared seventeen fuel blends by mixing ILP-17 with the fuels (Table 4), where HF-1 (pure IL-3) and HF-9 (pure IL-40) exhibited IDTs of >1000 ms and 392.5 ms, respectively. The optimal blend, HF-10, consisting of 3 wt.% ILP-17 and 97 wt.% IL-40, achieved a significantly shorter IDT of 25.3 ms while maintaining an I_{sp} of 240 s. Additionally, furfuryl alcohol was introduced to improve the solubility of ILP-17 in IL-3 and IL-40, but no notable reduction in IDT was observed. This study highlights the potential of sodium iodide as a promoter in developing green hypergolic bipropellant systems, offering a promising direction for environmentally friendly and high-performance propulsion technologies.

Table 4. IDTs of different hypergolic fuel (HF) combinations.

Fuels	IL-3 (wt.%)	IL-40 (wt.%)	Furfuryl Alcohol (wt.%)	$NaBH_3CN$ (wt.%)	ILP-17 (wt.%)	IDT (ms)
HF-1	100	0	0	0	0	>1000
HF-2	95	0	0	0	5	95.5
HF-3	93	0	0	0	7	93.6
HF-4	91	0	0	0	<9	75.5
HF-5	47.5	0	47.5	0	5	80.3
HF-6	46.5	0	46.5	0	7	64.5
HF-7	45.5	0	45.5	0	9	56.0
HF-8	44.5	0	44.5	0	<11	44.5
HF-9	0	100	0	0	0	392.5
HF-10	0	97	0	0	<3	25.3
HF-11	0	54	46	0	10	44.0
HF-12	0	76	19	0	5	82.3
HF-13	0	0	100	0	0	NO
HF-14	0	0	85	10	5	48.8
HF-15	0	0	83	10	7	34.0
HF-16	0	0	81	10	9	30.0
HF-17	0	0	79	10	<11	29.0

In 2021, Bhosale et al. developed two new hypergolic energetic copper (II) promoters, $[Cu^{II}(1\text{-H-imidazole})_4(BH_3CN)][BH_3CN]$ (ILP-18) and $[Cu^{II}(1\text{-methyl imidazole})_4(BH_3CN)_2]$ (ILP-19), for green propellant systems, with their synthetic pathway illustrated in Figure 13 [30]. These transition metal complexes demonstrated remarkable hypergolic reactivity with 95% H_2O_2, achieving IDTs of 3.75 ms and 8.50 ms for ILP-18 and ILP-19, respectively. To evaluate their practical application potential, ILP-18 was dissolved at 13 wt% in three distinct fuel systems: ionic liquid IL-3, tetraglyme (TG) solvent, and an equal-weight mixture of IL-3 and TG. Drop contact tests with 95% H_2O_2 revealed IDTs of 9.5 ms, 9.0 ms, and 7.8 ms for these systems, respectively, indicating a significant reduction compared to the pure ionic liquids and highlighting the superior performance of the IL-3/TG hybrid system. These findings underscore the promising potential of hypergolic ionic liquid fuel blends incorporating optimized amounts of energetic promoters and organic solvents as a viable alternative for green bipropellant fuels.

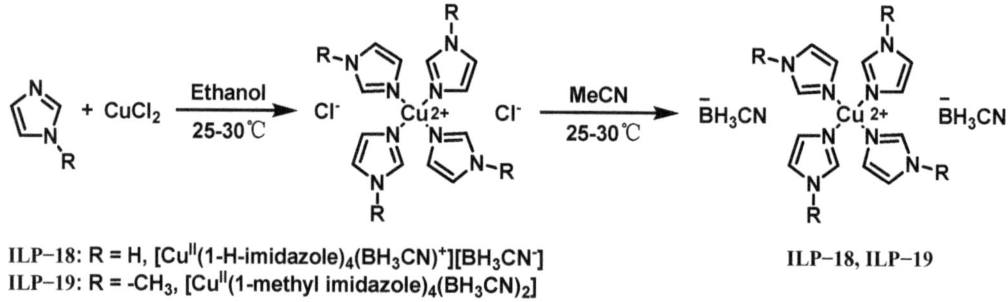

Figure 13. Synthesis of ILP-18 and ILP-19.

In 2024, to meet the performance requirements for rocket engine fire testing, Bhosale et al. investigated a novel hypergolic fuel blend designated as ILethCu01, composed of 9 wt% ILP-18 promoter in an equal-weight mixture of IL-3 (primary fuel) and ethanol (co-fuel) [31]. This ternary system demonstrated exceptional compatibility with 95 wt% H_2O_2, exhibiting a remarkably low viscosity of 3.85 mPa s and a favorable density of 0.90 g cm^{-3}. Theoretical performance calculations revealed an I_{sp} of 317 s and a ρI_{sp} of 403 s g cm^{-3}, surpassing conventional hypergolic systems. Crucially, drop-test experiments recorded an average IDT of 7.50 ms upon contact with 95 wt% H_2O_2, confirming rapid hypergolic initiation. These integrated properties—combining fluidity, energy density, and instantaneous ignition—establish ILethCu01/H_2O_2 as a leading candidate for next-generation green hypergolic propulsion systems.

In the same year, Seo et al. investigated the development of low-toxicity hypergolic propellants by enhancing triglyme-based fuels with ILP-18 and IL-3 [32]. Their study evaluated the IDTs of triglyme and triglyme–ionic liquid blends (TriGILs) containing varying ILP-18 concentrations, tested with 70, 90, and 95% H_2O_2. The 13Cu-TriGIL (TriGIL with 13% ILP-18) demonstrated exceptional performance, achieving an 8.0 ms IDT with 95% H_2O_2 compared to 18.7 ms for triglyme alone. The ionic liquid integration also improved cryogenic stability, reducing the blend's freezing point from −45 °C (pure triglyme) to below −80 °C. The optimized fuel exhibited favorable propulsion characteristics, including a density of 1.017 g cm^{-3} and viscosity of 26.42 mPa s. Theoretical calculations revealed that 13Cu-TriGIL with 95% H_2O_2 exhibited a 9.7% higher ρI_{sp} than MMH/NTO, despite a 2.1% lower I_{sp}. This work underscores triglyme–ionic liquid blends as promising eco-friendly alternatives to toxic hypergolic propellants, combining rapid ignition, enhanced thermal stability, and reduced hazards while maintaining competitive performance metrics.

In 2021, Zhao et al. designed, prepared, and comprehensively characterized six halogen-free energetic complexes (ILP-20 to ILP-22 and ILP-26 to ILP-28) with their synthesis scheme illustrated in Figure 14 [33]. These compounds exhibited high densities (1.180 to 1.373 g cm^{-3}) and robust thermal stability, displaying decomposition temperatures which ranged from 165.1 to 269.8 °C. Their catalytic performance was evaluated for the hypergolic reaction between ionic liquid IL-3 and 90% H_2O_2. Pure IL-3 exhibited an initial IDT exceeding 4000 ms with 90% H_2O_2, while the addition of 10 wt% promoter loading drastically reduced this parameter. Notably, ILP-20 and ILP-26 demonstrated superior catalytic performance, achieving IDTs of 37 ms and 31 ms, respectively. In 2022, as a continuation, Zhao et al. designed and comprehensively characterized six other halogen-free energetic complexes (ILP-23 to ILP-25 and ILP-29 to ILP-31) [34]. Their synthesis scheme is shown in Figure 14. Their densities ranged from 1.157 to 1.347 g cm^{-3}. Their thermal decomposition temperatures ranged from 169.6 to 255.9 °C. After adding 10 wt.% of prepared promoters to IL-3, the shortest IDT achieved was 94 ms.

These twelve complexes comprise two distinct categories: six borohydride-rich isomeric energetic complexes (ILP-20 to ILP-25) and six higher-density boron-free energetic complexes (ILP-26 to ILP-31). These materials share structural similarities while exhibiting differentiated ignition properties, providing an ideal platform for elucidating the catalytic ignition mechanisms between ionic liquids and high-concentration H_2O_2. The ignition process occurs through two sequential phases: H_2O_2 and the catalyst come into contact, and copper ions rapidly catalyze the decomposition of H_2O_2, generating highly reactive free radicals accompanied by significant exothermic release; this thermal energy then elevates the system temperature while the radicals initiate ionic liquid combustion, synergistically accelerating reaction kinetics to achieve spontaneous ignition. Notably, these halogen-free complexes demonstrate exceptional practical advantages through simple synthesis methods, high yields and superior performance metrics, positioning them as promising hypergolic promoters for green bipropellant systems that simultaneously address energy density requirements and environmental sustainability concerns.

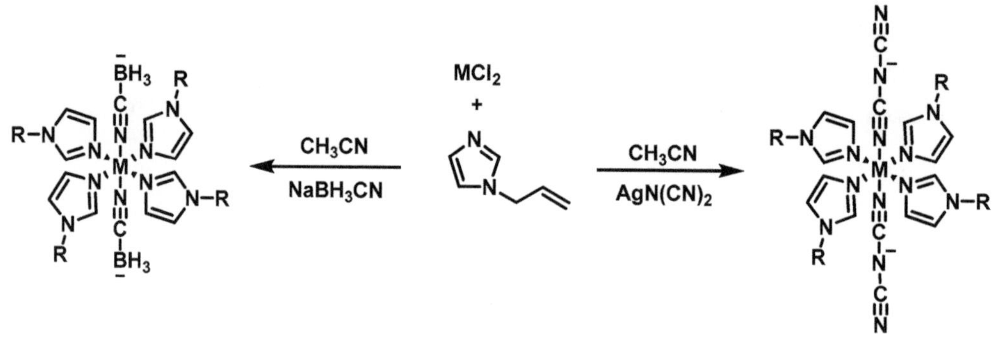

(ILP−20 to ILP−25)

(ILP−26 to ILP−31)

ILP−20 to ILP−22: R=allyl, M=Cu^{2+}, Ni^{2+}, Mn^{2+}
ILP−23 to ILP−25: R=ethyl, M=Cu^{2+}, Ni^{2+}, Mn^{2+}
ILP−26 to ILP−28: R=allyl, M=Cu^{2+}, Ni^{2+}, Mn^{2+}
ILP−29 to ILP−31: R=ethyl, M=Cu^{2+}, Ni^{2+}, Mn^{2+}

Figure 14. Synthesis of ILP-20 to ILP-31.

Few borohydride-rich ionic liquids exhibit spontaneous ignition with H_2O_2, necessitating the use of diverse promoters to enhance their hypergolic performance, as summarized in Table 5. Analysis reveals that these promoters primarily consist of metal elements (such as copper and iron), borides, and iodides, which reduce the Gibbs free energy barrier or catalyze substrate decomposition, thereby achieving lower IDTs. Notably, ILP-18 at 13 wt% concentration paired with IL-3 and 95% H_2O_2 delivered the highest performance, achieving the shortest IDT of 9.5 ms. ILethCu01 (9% ILP-18) achieves an ultralow IDT of 7.5 ms with a viscosity of 3.85 mPa·s. Promoter solubility in ionic liquids is critical, prompting strategies such as blending organic solvents into ionic liquid fuels to improve solubility. Experimental results confirm that solvent-optimized ionic liquid mixtures effectively reduce IDTs due to enhanced dissolution of active components. For instance, replacing pure IL-3 with an equal-weight blend of IL-3 and TG (an organic solvent) shortened the IDT from 9.5 ms to 7.8 ms, demonstrating the synergistic benefits of solubility-driven formulation design. Meanwhile, increased viscosity may impair combustion efficiency. The viscosity of IL-5 with ILP-14 rises from 28 to 47 mPa s. Future efforts should prioritize designing multifunctional promoters that balance catalytic activity with physicochemical properties, alongside eco-friendly and cost-effective synthesis routes.

Table 5. Physicochemical properties of borohydride-rich ionic liquid–H_2O_2 propellant systems with promoters.

IL	Promoter	w_{ILP} [a] (wt%)	ρ [b] (g cm^{-3})	η [c] (mPa s)	T_d [d] (°C)	T_m [e] (°C)	ΔH_f [f] (kJ mol^{-1})	I_{sp} [g] (s)	IDT [h] (ms)	Oxidizer
IL-3	ILP-1	8	3.13	-	-	-	-	237	69	70% H_2O_2
									24	95% H_2O_2
IL-3	ILP-2	8	2.73	-	-	-	-	237	45	70% H_2O_2
									17	95% H_2O_2
IL-3	NO	-	0.98	19	247	-	-	269	>1000	95% H_2O_2
IL-3	ILP-3	10	1.02	42	220	-	-	265	37	95% H_2O_2
IL-3	ILP-4	10	1.02	48	221	-	-	261	36	95% H_2O_2
IL-3	ILP-5	10	1.03	50	219	-	-	264	24	95% H_2O_2
IL-3	ILP-6	10	1.02	55	214	-	-	262	38	95% H_2O_2

Table 5. *Cont.*

IL	Promoter	w_{ILP} [a] (wt%)	ρ [b] (g cm^{-3})	η [c] (mPa s)	T_d [d] (°C)	T_m [e] (°C)	ΔH_f [f] (kJ mol^{-1})	I_{sp} [g] (s)	IDT [h] (ms)	Oxidizer
IL-40	NO	-	0.93	5	263			267	393	95% H_2O_2
IL-40	ILP-3	10	1.00	80	161	-	-	263	30	95% H_2O_2
IL-40	ILP-4	10	1.01	87	162	-	-	259	23	95% H_2O_2
IL-40	ILP-5	10	1.01	82	160	-	-	262	14	95% H_2O_2
IL-40	ILP-6	10	1.01	89	158	-	-	260	28	95% H_2O_2
IL-41	NO	-	1.11	35	203	<−70	359	-	NO	90% H_2O_2
IL-41	ILP-7	15	-	-	-	-	-	-	ignition	90% H_2O_2
IL-42	NO	-	1.13	52	193	<−70	472	-	NO	90% H_2O_2
IL-42	ILP-7	15	-	-	-	-	-	-	ignition	90% H_2O_2
IL-43	NO	-	1.12	28	241	<−70	242	-	NO	90% H_2O_2
IL-43	ILP-7	15	-	-	-	-	-	-	ignition	90% H_2O_2
IL-3	ILP-8	5	-	-	-	-	-	-	139	95% H_2O_2
IL-3	ILP-9	5	-	-	-	-	-	-	395	95% H_2O_2
IL-3	ILP-10	5	-	-	-	-	-	-	887	95% H_2O_2
IL-3	ILP-11	5	-	-	-	-	-	-	73	95% H_2O_2
IL-3	ILP-12	5	1.30	24	242	-	-	240	87	95% H_2O_2
IL-3	ILP-13	10	1.03	28	205	-	-	239	38	95% H_2O_2
IL-3	ILP-14	10	1.03	35	210	-	-	240	31	95% H_2O_2
IL-5	ILP-13	10	1.00	39	195	-	-	242	89	95% H_2O_2
IL-5	ILP-14	10	1.01	47	197	-	-	242	56	95% H_2O_2
IL-3	ILP-18	13	0.98	26	156	-	-	240	9.5	95% H_2O_2
TG	ILP-18	13	1.05	10	162	-	-	270	9.0	95% H_2O_2
IL-3/TG	ILP-18	13	1.02	21	160	-	-	248	7.8	95% H_2O_2
ILethCu01 (IL-3/ethanol)	ILP-18	9	0.90	4	-	-	-	317	7.5	95% H_2O_2
triglyme	ILP-18	13	1.02	5	156	−45	-	-	18.7	95% H_2O_2
TriGIL (IL-3/triglyme)	ILP-18	13	1.02	26	233	<−80	-	-	8.0	95% H_2O_2
IL-3	ILP-20	10	-	-	>240	-	-	278	37	90% H_2O_2
IL-3	ILP-22	10	-	-	>240	-	-	277	62	90% H_2O_2
IL-3	ILP-26	10	-	-	>240	-	-	277	31	90% H_2O_2
IL-3	ILP-28	10	-	-	>240	-	-	277	70	90% H_2O_2
IL-3	ILP-23	10	0.94	-	250	-	261	287	104	90% H_2O_2
IL-3	ILP-25	10	0.95	-	257	-	229	287	105	90% H_2O_2
IL-3	ILP-29	10	0.96	-	256	-	243	287	94	90% H_2O_2
IL-3	ILP-31	10	0.96	-	249	-	232	286	95	90% H_2O_2

[a] concentration of promoter (ILP). [b] density at 25 °C. [c] viscosity at 25 °C. [d] decomposition temperature. [e] melting temperature. [f] heat of formation. [g] specific impulse (ILP-12: under frozen flow conditions, P_c = 25 atm, P_e = 1 atm, A_e/A_t = 4; ILP-13 to ILP-14: under equilibrium conditions, P_c = 2.5 MPa; ILP-18 with IL-3: under frozen flow conditions, P_c = 25 atm, P_e = 1 atm, A_e/A_t = 4; ILP-18 with TG, IL-3/TG and ILethCu01: under frozen flow conditions, P_c = 1 MPa, A_e/A_t = 100; ILP-18 with triglyme and TriGIL: under frozen flow conditions, P_c = 9 bar, A_e/A_t = 150). [h] ignition delay time.

3.2. Promoters for ILs Based on Thiocyanate Anions and Hydrogen Peroxide

The previous text mentioned that Lauck et al. conducted drop tests with two different ionic liquids based on thiocyanate anions (IL-22 and IL-23) and 96.1% H_2O_2 in 2021 [21], reporting average IDTs of 31.7 ms and 45 ms, respectively. To optimize performance, copper thiocyanate (ILP-32) was introduced as a catalytic additive into IL-22. Promoted fuels containing different amounts of ILP-32 were systematically evaluated, revealing a non-linear relationship between additive concentration and ignition efficiency: the minimum delay of 13.9 ms was achieved at 5 wt% ILP-32, beyond which delays increased marginally. However, this enhancement came with trade-offs—fuel viscosity rose from 20 mPa s (pure IL-22) to 29.6 mPa s at 5 wt% ILP-32, while metallic additives introduced secondary challenges including I_{sp} degradation and combustion chamber fouling risks from particulate formation. These results highlight the critical balance required between ignition kinetics and operational practicality in catalytic hypergolic fuel design.

In the same year, Stützer et al. expanded the investigation of ILP-32's catalytic effects on the hypergolic reaction between IL-22 and 96.1% H_2O_2 by conducting laboratory-scale drop tests and flame emission spectroscopy to analyze combustion chemistry [35]. Their spectral analysis revealed that dissolving 5 wt% ILP-32 in IL-22 not only reduced the IDT (from 31 ms for pure IL-22 to 15 ms) but also elevated flame temperatures during rapid combustion. This dual functionality—enhancing both ignition kinetics and combustion intensity—demonstrates the critical role of ILP-32 in optimizing hypergolic performance. The study provides mechanistic insights into how transition metal additives modulate redox dynamics at the fuel–oxidizer interface, advancing the rational design of catalytic promoters for high-efficiency green propellants.

In 2022, Lauck et al. investigated the performance of a novel green hypergolic propellant, HIP_11, composed of 97% H_2O_2 as the oxidizer and a fuel blend of IL-22 with 5% ILP-32, aiming to evaluate the ignition reliability, combustion efficiency, and stability in a small "battleship" thruster designed for robustness and repeated testing [36]. Initial injector tests demonstrated that increasing the ILP-32 concentration from 1% to 5% markedly reduced IDTs to 13.9 ms, ensuring consistent hypergolic initiation. Subsequent hot-fire campaigns achieved stable combustion at chamber pressures exceeding 7 bar, delivering 93–98% combustion efficiency with exceptionally low pressure oscillations (1.1–1.7% of chamber pressure) and rapid rise times under 13 ms. This study marked the first successful demonstration of an ionic liquid–hydrogen peroxide system in operational thruster conditions, showing HIP_11's viability as a low-toxicity alternative to conventional propellants. The system's simplicity, repeatable ignition performance, and efficiency comparable to established hypergolic technologies highlight its potential for practical aerospace applications while addressing environmental and safety concerns.

Li et al. investigated the hypergolic ignition behavior of imidazolium thiocyanate-based ionic liquids blended with ethylene glycol (ILP-33) and propylene glycol (ILP-34) as additives in 90% H_2O_2 [37]. Their study revealed a counterintuitive trend: increasing the molar ratio of either glycol additive led to longer IDTs compared to the pure ionic liquid baseline. At fixed collision velocities, higher additive loadings further degraded combustion efficiency, with IDTs scaling inversely with glycol concentration. Notably, blends containing elevated ILP-33 or ILP-34 exhibited weakened flame propagation and reduced energy release rates, conclusively demonstrating that glycol additives impair, rather than enhance, the combustion performance of thiocyanate ionic liquids in concentrated H_2O_2 systems. This critical finding challenges conventional assumptions about oxygenated co-solvents, underscoring the need for alternative promoter strategies in hypergolic fuel design.

Park et al. investigated the impact of oxidizing additives, such as $LiNO_3$ (ILP-35) and NH_4NO_3 (ILP-36) [38], on the physical properties and ignition performance of green hypergolic systems composed of ionic liquid fuels (IL-22 and IL-23) and H_2O_2. While the additives negatively influenced the theoretical performance metrics, specifically reducing I_{sp} and ρI_{sp}, they significantly improved the low-temperature stability of the oxidizer mixtures. Notably, the addition of just 5 wt% ILP-35 to 95% H_2O_2 lowered the freezing point to −30 °C, demonstrating its exceptional efficacy. Furthermore, the oxidizing additives markedly enhanced ignition performance, with ILP-35 outperforming ILP-36 in drop tests. This improvement was amplified at lower H_2O_2 concentrations: although ILP-36 degraded ignition performance when combined with 90–95 wt% H_2O_2, the hypergolicity limit was extended when ILP-35 was incorporated into 60% H_2O_2 and paired with IL-23. These findings highlight the potential of nitrate salts as functional additives for tailoring H_2O_2-based oxidizers to achieve balanced cryogenic stability and ignition characteristics, despite their trade-offs in theoretical propulsion performance.

The physicochemical properties of thiocyanate-rich ionic liquid–H_2O_2 propellant systems with promoters are detailed in Table 6, revealing limited promoter diversity for thiocyanate-based formulations. Among the tested promoters (ILP-32 to ILP-36), inorganic salts (ILP-32, ILP-35 and ILP-36) effectively enhanced hypergolic ignition between thiocyanate-rich ionic liquids and high-concentration H_2O_2, whereas organic promoters (ILP-33 and ILP-34) exhibited that the addition of them correlated with prolonged IDT, disqualifying them for this propellant system. Notably, ILP-32 at a 5% concentration achieved the shortest IDT of 13.9 ms, outperforming the other two nitrates ILP-35 and ILP-36. Further analysis suggests that ILP-35 exhibits superior catalytic performance compared to ILP-36, likely due to its specific metal ion content, which accelerates reaction kinetics. These findings underscore the importance of continued exploration of inorganic salt promoters to optimize hypergolic performance in bisulfate ionic liquid–hydrogen peroxide systems, highlighting inorganic additives as a priority for future research.

Table 6. Physicochemical properties of thiocyanate-rich ionic liquid–H_2O_2 propellant systems with promoters.

IL	Promoter	w_{ILP} [a] (wt%)	ρ [b] (g cm^{-3})	η [c] (mPa s)	T_d [d] (°C)	T_m [e] (°C)	ΔH_f [f] (kJ mol^{-1})	I_{sp} [g] (s)	IDT [h] (ms)	Oxidizer
IL-22	ILP-32	5	1.15	30	273	-	-	-	13.9	96.1% H_2O_2
IL-22	NO	-	-	-	-	-	-	320	23.0	95% H_2O_2
IL-22	ILP-35	0.5	-	-	-	-	-	320	22.9	95% H_2O_2
IL-22	ILP-35	1	-	-	-	-	-	319	21.5	95% H_2O_2
IL-22	ILP-35	5	-	-	-	-	-	316	19.3	95% H_2O_2
IL-22	ILP-35	20	-	-	-	-	-	302	15.0	95% H_2O_2
IL-22	ILP-36	0.5	-	-	-	-	-	320	21.6	95% H_2O_2
IL-22	ILP-36	1	-	-	-	-	-	320	21.9	95% H_2O_2
IL-22	ILP-36	5	-	-	-	-	-	319	23.0	95% H_2O_2
IL-22	ILP-36	20	-	-	-	-	-	313	27.8	95% H_2O_2
IL-23	NO	-	-	-	-	-	-	322	35.1	95% H_2O_2
IL-23	ILP-35	0.5	-	-	-	-	-	322	33.1	95% H_2O_2
IL-23	ILP-35	1	-	-	-	-	-	321	32.2	95% H_2O_2
IL-23	ILP-35	5	-	-	-	-	-	318	30.1	95% H_2O_2
IL-23	ILP-35	20	-	-	-	-	-	303	23.7	95% H_2O_2
IL-23	ILP-36	0.5	-	-	-	-	-	320	34.6	95% H_2O_2
IL-23	ILP-36	1	-	-	-	-	-	322	32.7	95% H_2O_2
IL-23	ILP-36	5	-	-	-	-	-	320	31.6	95% H_2O_2
IL-23	ILP-36	20	-	-	-	-	-	314	36.7	95% H_2O_2

[a] concentration of promoter (ILP). [b] density at 25 °C. [c] viscosity at 25 °C. [d] decomposition temperature. [e] melting temperature. [f] heat of formation. [g] specific impulse (under equilibrium conditions, P_c = 2 MPa, A_e/A_t = 70). [h] ignition delay time.

3.3. Promoters for ILs Based on Dicyanamide Anions and Hydrogen Peroxide

In addition to borohydride-rich and thiocyanate-rich ionic liquids, dicyanamide anion-based ionic liquids are widely employed as propellant fuels. In 2014, Schneider et al. pioneered the use of metal-containing ionic liquids as hypergolic bipropellants to enhance ignition performance [39]. Their work focused on synthesizing 1-butyl-3-methylimidazolium chloroferrate ([BMIM][$FeCl_4$], ILP-37, Figure 15), which acted as a hypergolic promoter in reactions between ionic liquids and 98% H_2O_2. When tested with diaminomethylene azide dicyanamide ([DMAZ][$N(CN)_2$], IL-50, Figure 15) and H_2O_2, ILP-37 at a concentration of 8 wt% demonstrated its effectiveness as a hypergolic catalyst by significantly reducing IDTs to 110 ms with 90% H_2O_2 and 130 ms with 98% H_2O_2. This improvement stems from the ability of metal-containing ionic liquids to accelerate H_2O_2 decomposition, thereby enabling rapid ignition in bipropellant systems. These findings established a novel strategy for designing high-performance hypergolic formulations by integrating catalytic metal ions into ionic liquid frameworks, offering a pathway to optimize both reaction kinetics and system reliability.

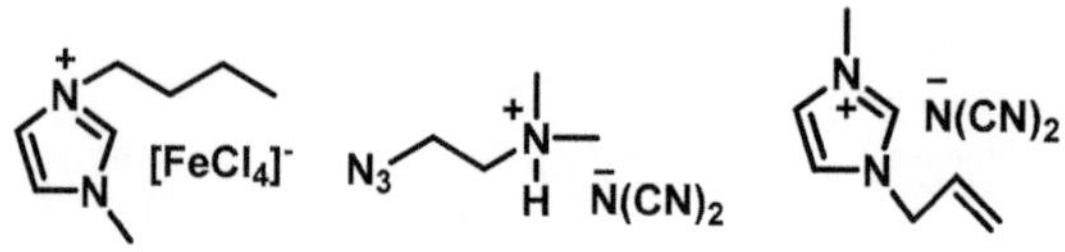

Figure 15. Structures of ILP-37 and IL-50.

In 2017, Weiser et al. conducted a study investigating the ignition performance of 1-allyl-3-methylimidazolium dicyanamide ([AMIM][N(CN)$_2$], IL-51, Figure 15) when combined with high-concentration H_2O_2 and a copper-based catalyst at a concentration of 15 wt% [40]. The team evaluated their ignition behavior through ignition tests using 95% H_2O_2, 90% H_2O_2, 80% H_2O_2, and 70% H_2O_2, which demonstrated remarkably short IDTs of 9 ms, 11 ms, 22 ms, and 66 ms, respectively. Recognizing the practical limitations of pure high-concentration H_2O_2, including long-term stability issues and low-temperature solidification, the researchers investigated additive-modified oxidizer formulations. They incorporated ammonium nitrate (AN, ILP-38), ammonium dinitramide (ADN, ILP-39), and urea peroxide (Urea, ILP-40) to improve thermal stability, lower melting points, and optimize oxygen balance and combustion efficiency. This investigation led to the development of 40 distinct oxidizer mixtures with varying mass ratios of H_2O_2, ILP-38, ILP-39 and ILP-40, all paired with IL-40 as the fuel component. The resulting IDTs across all formulations showed significant variation, spanning from 15 ms to nearly 200 ms, highlighting the critical role of oxidizer composition in tuning hypergolic performance. These findings underscore the exceptional reactivity of ionic liquid–oxidizer combinations when using high-purity H_2O_2, while simultaneously providing a framework for tailoring ignition characteristics through strategic additive selection and formulation optimization.

The previous text mentioned that Wang et al. synthesized promoters ILP-13 and ILP-14 in 2020, in order to investigate their ignition-enhancing effects on the ionic liquid fuels IL-3 and IL-10, while also evaluating their performance with 1-ethyl-3-methylimidazolium dicyanamide ([EMIM][N(CN)$_2$], IL-52) [28]. Their experiments revealed that blending IL-52 with 10% ILP-14 reduced the IDT to 42 ms when combined with 95% H_2O_2, whereas the same concentration of ILP-13 resulted in a slightly longer delay of 54 ms. However, the incorporation of these promoters introduced trade-offs: fuel mixtures exhibited marginally increased density and viscosity, with IL-52's inherently high base viscosity (69 mPa s) further compromising overall performance. Consequently, the modified IL-52 formulations underperformed compared to IL-3 and IL-5 in comprehensive evaluations, demonstrating that promoter-enhanced ignition efficiency must be balanced against detrimental physical property changes in hypergolic fuel design.

In 2022, Wang et al. provided a strategy for designing promoters by the synergy of cations and anions to seek green bipropellants [41]. They designed two difunctional promoters containing imidazolium-substituted borohydride cations and iodocuprate anions for hypergolic ignition of ionic liquids without BH_4^-/BH_3CN^- anions and 90% H_2O_2, which are [Bis(1-ethyl-1H-imidazole-3-ium-3-yl)dihydroboronium][Cu_3I_4] (ILP-41) and [Bis(1-allyl-1H-imidazole-3-ium-3-yl) dihydroboronium]$_2$[Cu_4I_6] (ILP-42). The synthesis scheme of ILP-41 and ILP-42 is shown in Figure 16. The study employed the ionic liquids [AMIM][N(CN)$_2$] (IL-40) and [BMIM][N(CN)$_2$] (IL-53) as fuels. Testing revealed that ILP-41 and ILP-42 had strong ignition-promoting potential. IL-40 with 5 wt% ILP-41 achieved an IDT of 75.0 ms, while increasing ILP-42 concentrations from 3 to 20 wt% progressively reduced IDTs, culminating in a 34.0 ms delay with 20 wt% ILP-42. For IL-53, 5 wt% additions of ILP-41 and ILP-42 yielded delays of 161 ms and 112.5 ms, respectively. Their efficacy stems from their dual-action mechanism: iodocuprate anions catalyze the exothermic decomposition of H_2O_2, while borohydride-functionalized cations act as ignition initiators. This work demonstrates how strategic cation–anion coordination in promoter design can advance environmentally sustainable hypergolic propellant systems.

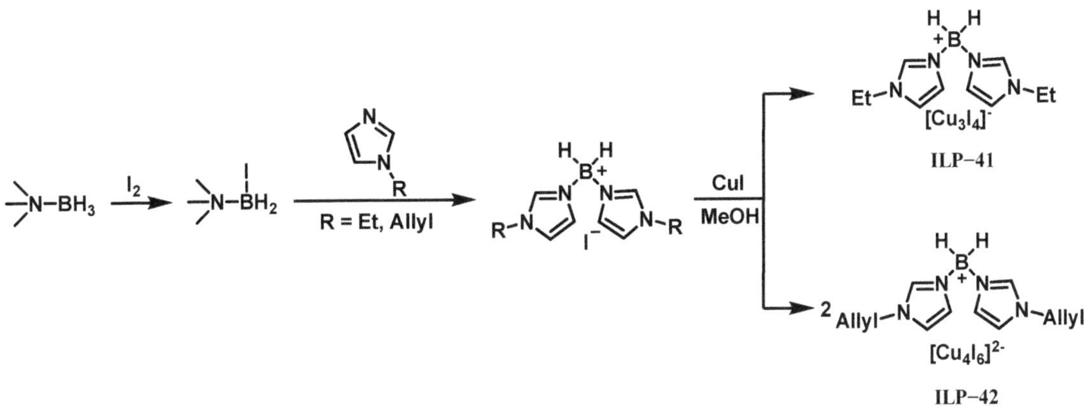

Figure 16. Synthesis of ILP-41 and ILP-42.

In 2024, Liao et al. introduced an innovative strategy to enhance hypergolic ignition in ionic liquid fuels using H_2O_2 as the oxidizer and copper-based ionic liquid promoters (ILP-43 to ILP-47) structurally tailored to mimic the cation composition of the fuel molecules (IL-40, IL-52 to IL-55), as illustrated in Figure 17 [42]. Leveraging the "like dissolves like" principle, these fuel–promoter pairs achieved full miscibility across most ratios, though ILP-47 exhibited slightly reduced solubility, requiring a 12% concentration for complete mixing with IL-52. Testing demonstrated a clear correlation between solubility and ignition performance: IL-52 paired with ILP-47 showed a prolonged delay of 59 ms, while combinations with the other four promoters all achieved sub-40 ms ignition times. ILP-43 emerged as the most effective promoter, enabling IL-52 to reach a remarkably short 16 ms delay. This superiority extended to other fuels, with all four ionic liquids paired with ILP-43 consistently igniting in under 40 ms when combined with 90% H_2O_2. The study underscores how strategic alignment of promoter–fuel molecular structures optimizes miscibility and ignition efficiency, advancing the development of high-performance hypergolic bipropellant systems.

In summary, the physicochemical properties of dicyanamide-rich ionic liquid–hydrogen peroxide propellant systems with promoters are shown in Table 7. Promoters for dicyanamide-rich ionic liquids paired with H_2O_2 have evolved to incorporate catalytic metal ions (such as iron and copper) and reactive functional groups (such as borohydride and iodocuprate) within their ionic frameworks. These promoters, such as ILP-37, ILP-41, and ILP-42, enhance ignition by dual mechanisms: metal anions catalyze the exothermic decomposition of H_2O_2, while functionalized cations initiate rapid fuel oxidation. Structural alignment between promoters (ILP-43 to ILP-47) and fuels (IL-51 to IL-55) further optimizes miscibility and ignition efficiency, achieving IDTs below 40 ms. Despite these advancements, inherent trade-offs persist between catalytic activity and physicochemical properties, necessitating continued innovation. Future efforts should focus on designing eco-friendly, multifunctional promoters with tailored cation–anion synergies, balanced physicochemical properties, and scalable synthesis routes to advance sustainable hypergolic bipropellant systems.

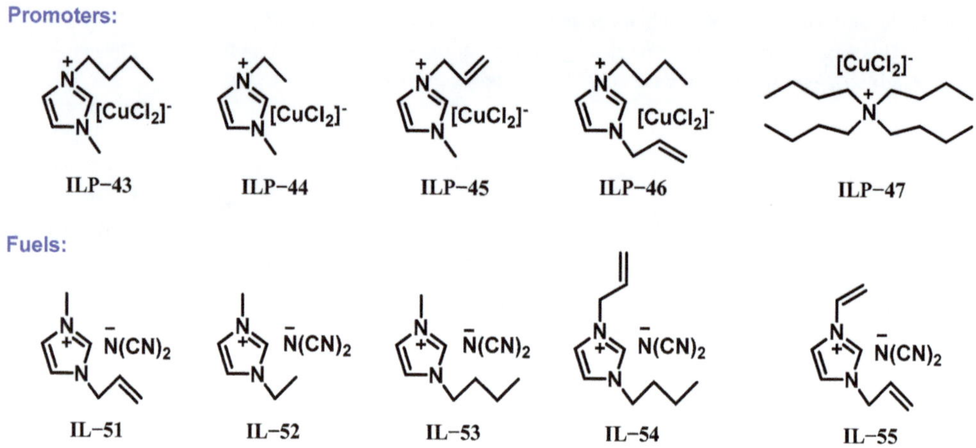

Figure 17. Structures of the promoters (ILP-43 to ILP-47) and the fuels (IL-51 to IL-55).

Table 7. Physicochemical properties of dicyanamide-rich ionic liquid–H_2O_2 propellant systems with promoters.

IL	Promoter	w_{ILP} [a] (wt%)	ρ [b] (g cm^{-3})	η [c] (mPa s)	T_d [d] (°C)	T_m [e] (°C)	ΔH_f [f] (kJ mol^{-1})	I_{sp} [g] (s)	IDT [h] (ms)	Oxidizer
IL-50	ILP-37	8	-	-	-	-	-	-	110	90% H_2O_2
IL-50	ILP-37	8	-	-	-	-	-	-	130	98% H_2O_2
IL-51	ILP	15	-	-	-	−5	-	-	9	95% H_2O_2
IL-51	ILP	15	-	-	-	−11	-	-	11	90% H_2O_2
IL-51	ILP	15	-	-	-	−25	-	-	22	80% H_2O_2
IL-51	ILP	15	-	-	-	−38	-	-	66	70% H_2O_2
IL-52	NO	-	1.12	69	205	-	-	235	>1000	95% H_2O_2
IL-52	ILP-13	10	1.14	78	182	-	-	235	54	95% H_2O_2
IL-52	ILP-14	10	1.15	84	188	-	-	235	42	95% H_2O_2
IL-40	ILP-41	5	1.14	17	257	<−70	-	272	75	90% H_2O_2
IL-40	ILP-42	3	1.13	19	256	<−70	-	275	112	90% H_2O_2
IL-40	ILP-42	5	1.14	17	254	<−70	-	274	74	90% H_2O_2
IL-40	ILP-42	10	1.15	24	252	<−70	-	272	59	90% H_2O_2
IL-40	ILP-42	15	1.20	31	248	<−70	-	270	47	90% H_2O_2
IL-40	ILP-42	20	1.25	40	244	<−70	-	268	34	90% H_2O_2
IL-53	ILP-41	5	1.09	27	273	<−70	-	264	161	90% H_2O_2
IL-53	ILP-42	5	1.09	27	272	<−70	-	266	113	90% H_2O_2
IL-52	ILP-43	12	1.13	35	238	-	-	297	16	90% H_2O_2
IL-52	ILP-44	12	1.15	37	240	-	-	297	36	90% H_2O_2
IL-52	ILP-45	12	1.15	35	236	-	-	298	29	90% H_2O_2
IL-52	ILP-46	12	1.13	34	238	-	-	296	20	90% H_2O_2
IL-52	ILP-47	12	1.11	40	239	-	-	299	59	90% H_2O_2
IL-51	ILP-43	12	1.13	42	239	-	-	296	26	90% H_2O_2
IL-53	ILP-43	12	1.10	52	243	-	-	295	36	90% H_2O_2
IL-54	ILP-43	12	1.17	40	212	-	-	298	27	90% H_2O_2
IL-55	ILP-43	12	1.16	43	214	-	-	298	17	90% H_2O_2

[a] concentration of promoter (ILP). [b] density at 25 °C. [c] viscosity at 25 °C. [d] decomposition temperature. [e] melting temperature. [f] heat of formation. [g] specific impulse (ILP-13 to ILP-14: under equilibrium conditions, P_c = 2.5 MPa; ILP-41 to ILP-42: using EXPLO5 with P_c and initial temperature as 7 MPa and 3300 K; ILP-43 to ILP-47: using NASA CEA with P_c and ambient pressure as 69.8×10^5 Pa and 1.0×10^5 Pa). [h] ignition delay time.

3.4. Promoters for ILs Based on Other Anions and Hydrogen Peroxide

The previous text mentioned that Schneider et al. synthesized the promoter [BMIM][$FeCl_4$] (ILP-37), which is used in the self-ignition reaction between ionic liquids and 98% H_2O_2 in 2014 [39]. In fact, in addition to the ionic liquids [DMAZ][$N(CN)_2$] (IL-50), ILP-37 also promoted hypergolic ignition between many other ionic liquids such as 1-butyl-3-methylimidazole azide ([BMIM][N_3], IL-56), 2-hydroxyethylhydrazine nitrate ([HEH][NO_3], IL-57) and dimethylethylene azide trifluoroacetic acid ammonium ([DMAZ][TFA], IL-58), and 98% H_2O_2. The IDTs were 170 ms, 50 ms, and 960 ms, respectively.

In 2019, addressing challenges posed by the high viscosity and poor miscibility of 1-butyl-3-methylimidazolium acetate ([BMIM][Ac], IL-59) with H_2O_2, Lauck et al. developed modified fuel blends by incorporating organic solvents [43]. Two formulations were tested: an 81.3% IL-59 mixture with 10% ethanol (ILP-48) and 8.7% MAT (ILP-49), and a 72.8% IL-59 composition with 19.4% ILP-48 and 7.8% ILP-49. Both mixtures significantly reduced viscosity and achieved full miscibility with H_2O_2, enabling reliable ignition in drop tests with consistent delay times averaging approximately 28 ms. However, this solvent-based approach partially compromised key advantages of ionic liquids—specifically their inherently low vapor pressures and high densities—highlighting the need to explore alternative ionic liquids that retain these beneficial properties while improving compatibility with oxidizers. The study underscores the delicate balance required between optimizing fuel–oxidizer interactions and preserving the intrinsic advantages of ionic liquid propellants in hypergolic system design.

To conclude, the physicochemical properties of other ionic liquid–hydrogen peroxide propellant systems with promoters are shown in Table 8. Notably, metal-ion promoters (such as ILP-37) have been extended to non-conventional ionic liquids (such as IL-56/57/58), but their performance exhibits significant variability. For instance, IDTs range from 50 ms for IL-57 to 960 ms for IL-58, underscoring the critical role of ionic liquid selection in system optimization. Solvent blending strategies, exemplified by combinations like IL-59+ILP-48/49, effectively reduce viscosity (from 97 to 37 mPa s), thereby improving fluidity.

Table 8. Physicochemical properties of other ionic liquid–H_2O_2 propellant systems with promoters.

IL	Promoter	w_{ILP} [a] (wt%)	ρ [b] (g cm^{-3})	η [c] (mPa s)	T_d [d] (°C)	T_m [e] (°C)	ΔH_f [f] (kJ mol^{-1})	I_{sp} [g] (s)	IDT [h] (ms)	Oxidizer
IL-56	ILP-37	14	-	-	-	-	-	-	170	98% H_2O_2
IL-57	ILP-37	22	-	-	-	-	-	-	50	98% H_2O_2
IL-58	ILP-37	20	-	-	-	-	-	-	960	98% H_2O_2
IL-59	ILP-48/49	10/8.7	1.05	97	-	-	-	326	28	97% H_2O_2
IL-59	ILP-48/49	19.4/7.8	1.02	37	-	-	-	326	28	97% H_2O_2

[a] concentration of promoter (ILP). [b] density at 25 °C. [c] viscosity at 25 °C. [d] decomposition temperature. [e] melting temperature. [f] heat of formation. [g] specific impulse (under frozen flow conditions, P_c = 10 bar, A_e/A_t = 330). [h] ignition delay time.

4. Conclusions

In this paper, we review the recent trends and developments in green liquid propellants based on ionic liquid–hydrogen peroxide combinations and analyze in detail the synthesis and ignition performances of both self-igniting propellants and promoter-dependent propellants. The latest research has demonstrated that introducing functional promoters is a feasible strategy, which not only enables the rapid ignition of ionic liquids with high-concentration H_2O_2 but also significantly reduces IDTs and enhances the performance parameters of the ionic liquids.

The viscosity, I_{sp}, melting point, stability, solubility, and cost of promoters are critical determinants of their practical applicability. First, viscosity directly influences the propellant fluidity, which governs injection efficiency and combustion performance. For example, adding ILP-32 to IL-22 elevates the viscosity from 20 to 29.6 mPa s, while blending IL-52 with ILP-13 or ILP-14 increases the viscosity from 69 to 78 and 84 mPa s, respectively. Elevated viscosity raises flow resistance, impairing injection efficiency, and poor mixing due to high viscosity can lead to incomplete combustion. To address this, solvent blending or the design of low-viscosity promoters ($\eta < 50$ mPa s) is essential. I_{sp} is an important index to measure the performance of propellants. Higher I_{sp} (>300 s) can enhance rocket engine efficiency, payload capacity, and velocity. Low melting points (T_m < room temperature) ensure propellant liquidity under cryogenic conditions, preventing solidification and ensuring reliable supply. Thermal stability (T_d > 200 °C) is vital for safe storage and consistent ignition in engines. The solubility of solid promoters is also a critical parameter in fuel formulation design. These promoters must be uniformly dispersed within the fuel to form a homogeneous system, ensuring sufficient contact with the fuel and enabling effective catalytic action. However, low solubility can lead to phase separation or sedimentation, resulting in uneven local concentrations. This imbalance may cause delayed or incomplete combustion, ultimately reducing specific impulse and compromising reliability. Cost considerations are equally critical: metal-containing promoters like ILP-41 offer superior catalytic efficiency but require complex synthesis and expensive metals, whereas metal-free alternatives like ILP-32 are cost-effective but less active. Future advancements must balance catalytic performance, economic viability, and environmental sustainability.

Among the promoters analyzed, ILP-18 emerges as the most promising candidate for rocket propulsion. As shown in Table 5, the IL-3/ethanol blend (ILethCu01) with 9% ILP-18 achieves an ultralow IDT of 7.5 ms, a low viscosity of 3.85 mPa s, and a high I_{sp} of 317 s. Notably, ILP-18 retains reliable hypergolic ignition even at extreme temperatures (the TriGIL system can operate at −80 °C). This catalyst functions through two complementary mechanisms: copper ions accelerate the decomposition of hydrogen peroxide, while borohydrides drive rapid redox reactions. These processes enable fast ignition and sustained combustion, meeting the practical requirements of rocket engines. These attributes position ILP-18 as the leading green propellant candidate for next-generation propulsion systems.

However, the exploration in this field is still insufficient, and more in-depth research should be carried out in the following aspects in the future. Firstly, the library of ionic liquids should be expanded by designing different cations and anions and identifying their structure–ignition relationships. Secondly, the library of promoters should be expanded and the connection between structure and catalytic performance explored. In conclusion, the development of new ionic liquid–hydrogen peroxide combinations will undoubtedly continue to advance the field of environmentally friendly chemical propulsion. In the future, we can expect a wide range of researchers to develop more novel high-performance ionic liquids and promoters, further driving progress in this area.

Author Contributions: Conceptualization, D.Y. and Q.Z.; methodology, D.Y.; validation, Y.Z. and X.Z.; writing—original draft preparation, Y.Z. and X.Z.; resources, D.Y.; investigation, Y.Z.; writing—review and editing, Y.Z. and X.Z.; visualization, Y.Z.; supervision, Y.Z.; funding acquisition, Q.Z. All authors have read and agreed to the published version of the manuscript.

Funding: This research received no external funding.

Institutional Review Board Statement: Not applicable.

Informed Consent Statement: Not applicable.

Data Availability Statement: This article is a review of previous works, no new data were created. The related data can be found in the references.

Conflicts of Interest: The authors declare no conflicts of interest.

References

1. Kamal, F.; Yann, B.; Rachid, B.; Charles, K. Application of ionic liquids to space propulsion. *Appl. Ion. Liq. Sci. Technol.* **2011**, *447*, 466.
2. Li, Y.; Yang, Y.; Yi, Z.; Song, D.; Cheng, Y.; Li, Y. Sodium Azotetrazolate: A Novel Environmental-friendly Hydrogen-free Gas-Generating Pyrotechnics. *Chem. Eng. J.* **2021**, *413*, 127442. [CrossRef]
3. Zhang, D.; Zhang, P.; Yuan, Y.; Zhang, T. Hypergolic ignition by head-on collision of N, N, N′, N′-tetramethylethylenediamine and white fuming nitric acid droplets. *Combust. Flame* **2016**, *173*, 276–287. [CrossRef]
4. Zhang, X.; Yang, Z.; Nie, F.; Yan, Q. Recent advances on the crystallization engineering of energetic materials. *Energetic Mater. Front.* **2020**, *1*, 141–156. [CrossRef]
5. Bhosale, V.K.; Jeong, J.; Choi, J.; Churchill, D.G.; Lee, Y.; Kwon, S. Additive-promoted hypergolic ignition of ionic liquid with hydrogen peroxide. *Combust. Flame* **2020**, *214*, 426–436. [CrossRef]
6. Kapusta, Ł.J.; Boruc, Ł.; Kindracki, J. Pressure and temperature effect on hypergolic ignition delay of triglyme-based fuel with hydrogen peroxide. *Fuel* **2021**, *287*, 119370. [CrossRef]
7. Fu, L.; Jiao, B.; Wang, B.; Cui, S.; Hao, Y. Review on Boronium-Anion-Based Hypergolic Ionic Liquids. *Chin. J. Energetic Mater.* **2022**, *30*, 1165–1176.
8. Catoire, L.; Chaumeix, N.; Paillard, C. Chemical kinetic model for monomethylhydrazine/nitrogen tetroxide gas phase combustion and hypergolic ignition. *J. Propuls. Power* **2004**, *20*, 87–92. [CrossRef]
9. Frank, I.; Hammerl, A.; Klapötke, T.M.; Nonnenberg, C.; Zewen, H. Processes during the hypergolic ignition between monomethylhydrazine (MMH) and dinitrogen tetroxide (N2O4) in rocket engines. *Propellants Explos. Pyrotech. Int. J. Deal. Sci. Technol. Asp. Energetic Mater.* **2005**, *30*, 44–52. [CrossRef]
10. Nonnenberg, C.; Frank, I.; Klapötke, T.M. Ultrafast cold reactions in the bipropellant monomethylhydrazine/nitrogen tetroxide: CPMD simulations. *Angew. Chem. Int. Ed.* **2004**, *43*, 4586–4589. [CrossRef]
11. Thomas, A.; Stober, K.J.; Al Otaibi, R.; Alotaibi, M.; Almuqati, N.; Evans, B.J.; Gao, H.; Shreeve, J.N.; Cantwell, B.J. Ignition delay testing of various hypergolic ionic liquids and oxidizers. In Proceedings of the 52nd AIAA/SAE/ASEE Joint Propulsion Conference, Salt Lake City, UT, USA, 25–27 July 2016; p. 4782.
12. Zhang, Q.; Shreeve, J.n.M. Energetic ionic liquids as explosives and propellant fuels: A new journey of ionic liquid chemistry. *Chem. Rev.* **2014**, *114*, 10527–10574. [CrossRef] [PubMed]
13. Lauck, F.; Witte, J.; Negri, M.; Freudenmann, D.; Schlechtriem, S. Design and first results of an injector test setup for green hypergolic propellants. In Proceedings of the AIAA Propulsion and Energy 2019 Forum, Indianapolis, IN, USA, 19–22 August 2019; p. 4279.
14. Li, X.; Huo, H.; Li, H.; Nie, F.; Yin, H.; Chen, F.-X. Cyanotetrazolylborohydride (CTB) anion-based ionic liquids with low viscosity and high energy capacity as ultrafast-igniting hypergolic fuels. *Chem. Commun.* **2017**, *53*, 8300–8303. [CrossRef] [PubMed]
15. Ventura, M. Long term storability of hydrogen peroxide. In Proceedings of the 41st AIAA/ASME/SAE/ASEE Joint Propulsion Conference & Exhibit, Tucson, AZ, USA, 10–13 July 2005; p. 4551.
16. Sun, C.; Tang, S.; Zhang, X. Role of cation structures for energetic performance of hypergolic ionic liquids. *Energy Fuels* **2017**, *31*, 10055–10059. [CrossRef]
17. Schneider, S.; Hawkins, T.; Ahmed, Y.; Rosander, M.; Hudgens, L.; Mills, J. Green bipropellants: Hydrogen-rich ionic liquids that are hypergolic with hydrogen peroxide. *Angew. Chem. Int. Ed.* **2011**, *50*, 5886–5888. [CrossRef] [PubMed]
18. Bhosale, V.K.; Kulkarni, S.G.; Kulkarni, P.S. Theoretical performance evaluation of hypergolic ionic liquid fuels with storable oxidizers. *New J. Chem.* **2017**, *41*, 9889–9896. [CrossRef]
19. Bhosale, V.K.; Jeong, J.; Kwon, S. Ignition of boron-based green hypergolic fuels with hydrogen peroxide. *Fuel* **2019**, *255*, 115729. [CrossRef]
20. Wang, Z.; Fei, L.; Xia, H.; Jin, Y.; Zhang, Q. Organic superbase-mediated synthesis of borohydride ionic liquids as novel composite hypergolic fuels. *Energetic Mater. Front.* **2023**, *4*, 77–84. [CrossRef]
21. Lauck, F.; Balkenhohl, J.; Negri, M.; Freudenmann, D.; Schlechtriem, S. Green bipropellant development—A study on the hypergolicity of imidazole thiocyanate ionic liquids with hydrogen peroxide in an automated drop test setup. *Combust. Flame* **2021**, *226*, 87–97. [CrossRef]
22. Ricker, S.C.; Freudenmann, D.; Schlechtriem, S. The impact of cation structures on hypergolicity of thiocyanate ionic liquids with hydrogen peroxide. *Energy Fuels* **2021**, *35*, 16128–16133. [CrossRef]
23. Ricker, S.C.; Brüggemann, D.; Freudenmann, D.; Ricker, R.; Schlechtriem, S. Protic thiocyanate ionic liquids as fuels for hypergolic bipropellants with hydrogen peroxide. *Fuel* **2022**, *328*, 125290. [CrossRef]
24. Stölzle, S.C.; Kruse, L.; Freudenmann, D. Trialkylsulfonium Thiocyanate Ionic Liquids: Investigation on Temperature-Dependent Ignition Behavior of Green Hypergolic Propellants. *Propellants Explos. Pyrotech.* **2024**, *49*, e202400151. [CrossRef]

25. Chinnam, A.K.; Petrutik, N.; Wang, K.; Shlomovich, A.; Shamis, O.; Tov, D.S.; Sućeska, M.; Yan, Q.-L.; Dobrovetsky, R.; Gozin, M. Effects of closo-icosahedral periododborane salts on hypergolic reactions of 70% H_2O_2 with energetic ionic liquids. *J. Mater. Chem. A* **2018**, *6*, 19989–19997. [CrossRef]
26. Wang, K.; Chinnam, A.K.; Petrutik, N.; Komarala, E.P.; Zhang, Q.; Yan, Q.-L.; Dobrovetsky, R.; Gozin, M. Iodocuprate-containing ionic liquids as promoters for green propulsion. *J. Mater. Chem. A* **2018**, *6*, 22819–22829. [CrossRef]
27. Wang, Z.; Jin, Y.; Zhang, W.; Wang, B.; Liu, T.; Zhang, J.; Zhang, Q. Synthesis and hypergolic properties of flammable ionic liquids based on the cyano (1 H-1, 2, 3-triazole-1-yl) dihydroborate anion. *Dalton Trans.* **2019**, *48*, 6198–6204. [CrossRef]
28. Wang, K.; Liu, T.; Jin, Y.; Huang, S.; Petrutik, N.; Shem-Tov, D.; Yan, Q.-L.; Gozin, M.; Zhang, Q. "Tandem-action" ferrocenyl iodocuprates promoting low temperature hypergolic ignitions of "green" EIL–H_2O_2 bipropellants. *J. Mater. Chem. A* **2020**, *8*, 14661–14670. [CrossRef]
29. Bhosale, V.K.; Kim, K.-S.; Kwon, S.; Churchill, D.G. Sodium iodide: A trigger for hypergolic ignition of non-toxic fuels with hydrogen peroxide. In Proceedings of the AIAA Propulsion and Energy 2020 Forum, Virtually, 24–28 August 2020; p. 3824.
30. Bhosale, V.K.; Gwak, J.; Kim, K.-S.; Churchill, D.G.; Lee, Y.; Kwon, S. Rapid ignition of "green" bipropellants enlisting hypergolic copper (II) promoter-in-fuel. *Fuel* **2021**, *297*, 120734. [CrossRef]
31. Bhosale, V.K.; Lee, K.; Yoon, H.; Kwon, S. Green bipropellant: Performance evaluation of hypergolic ionic liquid-biofuel with hydrogen peroxide. *Fuel* **2024**, *376*, 132688. [CrossRef]
32. Seo, M.; Bhosale, V.K.; Im, H.; Kwon, S. Performance improvement of triglyme-based fuels using an ionic liquid with hydrogen peroxide. *Combust. Flame* **2024**, *270*, 113719. [CrossRef]
33. Zhao, X.; Wang, Z.; Qi, X.; Song, S.; Huang, S.; Wang, K.; Zhang, Q. Hunting for energetic complexes as hypergolic promoters for green propellants using hydrogen peroxide as oxidizer. *Inorg. Chem.* **2021**, *60*, 17033–17039. [CrossRef]
34. Zhao, X.; Wang, J.; Jin, Y.; Wang, K.; Zhang, Q. Energetic complexes as promoters for the green hypergolic bipropellant of EIL-H_2O_2 combinations. *FirePhysChem* **2022**, *2*, 185–190. [CrossRef]
35. Stützer, R.G.; Balkenhohl, J.; Lauck, F.; Oschwald, M.; Schlechtriem, S. Hypergolic reaction between green ionic liquid EMIM SCN and hydrogen peroxide in lab-scale drop test chamber. *Int. J. Energetic Mater. Chem. Propuls.* **2022**, *21*, 87–100. [CrossRef]
36. Negri, M.; Lauck, F. Hot Firing Tests of a Novel Green Hypergolic Propellant in a Thruster. *J. Propuls. Power* **2022**, *38*, 467–477. [CrossRef]
37. Li, Y.; Pang, Y.; Gao, Z.; Tan, Z.; Chen, P.; Tan, Y.; Hong, L. Study on the Hypergolic Characteristics of Imidazole Thiocyanate Ionic Liquids and Hydrogen Peroxide. *J. Xi'an JiaoTong Univ.* **2022**, *56*, 12–20.
38. Park, S.; Lee, K.; Kang, H.; Park, Y.; Lee, J. Effects of oxidizing additives on the physical properties and ignition performance of hydrogen peroxide-based hypergolic propellants. *Acta Astronaut.* **2022**, *200*, 48–55. [CrossRef]
39. Schneider, S.; Hawkins, T.W.; Ahmed, Y.; Rosander, M. Catalytic Hypergolic Bipropellants. US 8,758,531 B1, 24 June 2014.
40. Weiser, V.; Hürttlen, J.; Schaller, U.; Imiolek, A.; Kelzenberg, S. Green liquid oxidizers basing on solutions of ADN and AN in hydrogen peroxide for hypergolic propellants with high performance. In Proceedings of the 7th European Conference for Aeronautics and Space Sciences (EUCASS), Milan, Italy, 3–6 July 2017.
41. Wang, B.; Wang, Z.; Jin, Y.; Wang, K. Designing difunctional promoters for hypergolic ignitions of green bipropellants combining ionic liquids with H_2O_2. *Ind. Eng. Chem. Res.* **2022**, *61*, 17433–17439. [CrossRef]
42. Liao, S.; Liu, T.; Zhou, Z.; Wang, K.; Zhang, Q. Auto-ignition of ionic liquid fuels with hydrogen peroxide triggered by copper-containing liquid promoter. *Energetic Mater. Front.* **2024**, *5*, 41–46. [CrossRef]
43. Lauck, F.; Negri, M.; Freudenmann, D.; Schlechtriem, S. Study on hypergolic ignition of ionic liquid solutions. In Proceedings of the 8th European Conference for Aeronautics and Space Sciences, Madrid, Spain, 1–4 July 2019.

MDPI AG
Grosspeteranlage 5
4052 Basel
Switzerland
Tel.: +41 61 683 77 34

Molecules Editorial Office
E-mail: molecules@mdpi.com
www.mdpi.com/journal/molecules

www.ingramcontent.com/pod-product-compliance
Lightning Source LLC
Chambersburg PA
CBHW062102220926
51850CB00010B/2858